走向未来建筑

郭小平 编著

华中科技大学出版社
http://www.hustp.com
中国·武汉

序言

本书共16章，分别从不同方面讲述未来建筑创作，提出了值得建筑师们思考的课题。

作者依据进化的基本原理和生态学的理论，分析了人、建筑与自然生态之间的关系，并从能源和人类资源的角度论述了建筑的本质及现代科技在建筑中的应用的重要性，从而提醒建筑师在建筑创作中承担起保护自然与生态的责任。

在建筑的创作应重视光和空气的要素，并采用科技手法有效利用。建筑产业化和现代建筑的外围护墙的新技术的应用大大提高了建筑品质，是解决社会日益关注的能源节约和能源开发问题的有效途径，并使社会得以持续发展。

交流所涉及的公共空间和私有空间为和谐社会提供了必要条件，美和个性使建筑充满生机，使人们得到极大的精神享受。

作者在每章提供了相应的案例，这些案例主要来源于德国著名建筑师事务所克里斯多夫·英恩霍文（Ingenhoven and Partner）的实践作品，这些建筑充分反映了作者书中提出的创作理念。由于作者在英恩霍文建筑师事务所从事建筑创作共8年，并参与了多个项目的创作实践，因此，他对项目案例的创作理念及解析更具说服力。

这是一本值得建筑师阅读的书。

2009.5.20.

Preface

This book has 16 chapters, with content showing the impact on the future of architectural design from different aspects, proposing subjects well worth to consider for those who are engaged in this creative work.

From the basic principles of evolution and ecology theory, the author analyzed the relationship between people, building and the natural ecosystems. In view of energy and human resources, the author discussed the nature of the building and the importance of the application of modern science and technology in the construction, thus reminding architects to assume the responsibility for the protection of the nature and the ecosystems in their architectural creation.

The elements of light and air are being emphasized, and the effective use of modern technology is taken in today's architectural design. The industrialization of construction and the application of new technologies in modern architecture greatly improved the quality of building, seeking the solution of energy conservation and energy development which got increasingly widespread concern, thus promoting the sustainable social development.

The communication involved in both public and private spaces provided the necessary condition for a harmonious society; beauty and personality build full of vigor, so that people get a great spiritual enjoyment.

In each chapter, the author provided appropriate cases, mainly from the practical works by the famous German architect firm Christopher Ingenhoven (Ingenhoven & Partner). These buildings fully reflect the author's design philosophy. As the author had been engaged with architectural design at Ingenhoven Architects for eight years and participated in multiple projects, his creative concepts and resolutions are more convincingly depicted in this book.

This is a book worth reading for any architect.

Guanzhang Wu
2009.05.20

目录 /Table of Content

1 进化 EVOLUTION — 8
项目解读：重庆自然博物馆新馆 — 14
Case Study: Natural History Museum, Chongqing
项目解读：贵阳医学院图书馆 — 22
Case Study: Guiyang Medical Institution Library, Guiyang

2 生态学 ECOLOGY — 28
项目解读：德国埃森莱茵集团总部大楼 — 34
Case Study: Headquarter Building of RWE, Essen, Germany

3 能源 ENERGY — 48
项目解读：德国法兰克福商业银行 — 54
Case Study: Commerzbank, Frankfurt, Germany

4 城市与城市发展 CITY AND CITY DEVELOPMENT — 58
项目解读：德国杜塞尔多夫北城区CBD中心规划 — 64
Case Study: CBD Dusseldorf, Germany

5 简单与逻辑 SIMPLICITY AND LOGIC — 68
项目解读：韩国KIM HAE体育中心 — 74
Case Study: KIM HAE Stadium, South Korea
项目解读：宁波富春酒店 — 78
Case Study: Fuchun Hotel, Ningbo

6 光 LIGHT — 84
项目解读：上海科技馆 — 90
Case Study: Shanghai Science Technology Museum

7 空气 AIR — 96
项目解读：合肥招商银行大厦 — 102
Case Study: Merchants Bank Building, Hefei
项目解读：德国慕尼黑UPTOWN建筑群 — 106
Case Study: Uptown Munich, Germany

8 集合作用 SYNERGY — 112
项目解读：湖北省输变电工程公司办公楼 — 116
Case Study: Office Building of Hubei Power Transmission Company, Wuhan

9 工业产品与建筑 INDUSTRIAL PRODUCTS AND ARCHITECTURE   122
项目解读：法兰克福、巴黎、东京、底特律、日内瓦国际车展奥迪场馆   128
Case Study: Audi AG Pavilion, Frankfurt/Tokyo/Detroit/Geneva/Paris

10 外墙技术 FACADE TECHNOLOGY   136
项目解读：上海世茂国际广场   142
Case Study: Shimao International Plaza, Shanghai

11 交流 COMMUNICATION   148
项目解读：德国法兰克福汉沙航空公司总部办公楼   152
Case Study: Lufthansa AG Headquarter Building, Frankfurt, Germany

12 体量和外表 VOLUME AND SURFACE   166
项目解读：德国埃森火车站   170
Case Study: Railway Station, Essen, Germany
项目解读：兰州盛达金城广场城市综合体   174
Case Study: Shengda Jincheng Plaza City Complex, Lanzhou

13 效率 EFFICIENCY   182
项目解读：海口西海岸假日酒店   186
Case Study: Haikou Westcoast Holiday Inn, Haikou

14 个性 PERSONALITY   192
项目解读：合肥MV广场   196
Case Study: MV Plaza, Hefei
项目解读：兰州盛达金城广场城市综合体   202
Case Study: Shengda Jincheng Plaza City Complex, Lanzhou

15 美 BEAUTY   206
项目解读：德国汉堡空客A380组装大厅   210
Case Study: Airbus A380 Assembly Hall, Hamburg, Germany

16 未来建筑——零能源建筑 FUTURE ARCHITECTURE - ZERO-ENERGY BUILDING   218
项目解读：德国斯图加特21世纪斯图加特火车站   222
Case Study: 21st Century Train Station, Stuttgart, Germany

2 进化
EVOLUTION

生物和植物的进化过程可以成为新一代建筑师的指导，它的演变历程可以启迪建筑师进行创作。如果我们想做出成绩，就必须长期学习匹配、兼容、发展的进化系统。

THE NEW GENERATION OF ARCHITECTS MAY TAKE ENDLESS INSPIRATIONS FROM THE EVOLUTION OF LIFE, STRIVING FOR ADAPTATION, MATCHING, COMPATIBILITY, OPTIMIZATION AND PROGRESSION INDEFINITELY.

1 进化
EVOLUTION

1. 乌尔内伯登,一个瑞士的居住小区,位于1400m的海拔高度,使用原始的木材及传统的木结构建筑方式建造房屋,它们已有700年的历史

1. Urner Boden, a residential district in Switzerland, is located 1400m altitude. Houses are more than 700 years old, built with the original timber using the traditional wooden structure as construction method

"进化"一词源于拉丁语"evolvere"。进化是发展的意思,是指一个族群的基因库为了回应环境压力、自然选择和基因突变而逐渐(经过几百万年)产生改变的过程。所有的生命形式都会经历这个过程。

进化过程

在科学实验的模型中,我们将进化过程简单化,如同一个连续的链条或盘旋上升的组件一样简单。因外界的影响产生的变化将稳定的进化状态演化为不稳定的状态,然后在组建过程中引发新的结构产生,并取代了旧的结构形式,最终将形成全新及稳定的状态,这一状态在新的循环系统中稳定发展和交融。因此,可以推断,进化系统在寻求稳定的、受争议的状态的同时更新了自己的内部组织,在原则上,大量的数据都表明了这一状态存在着内部相互竞争并相互满足。

在生物学的进化过程中,使自我的界限达到一个稳定的状态已成为可能,这一稳定的环境决定性地影响了转换(进化)过程。

滑雪

新的技术使竞技运动有了革命性的发展,同时加快了新技术对体育的应用,通过辅助的设备、竞技技术不断地打破世界纪录。这一结果有一个逻辑的发展过程:自起步、运动到结束,包含了力量、能量、技术、简洁和自然等多种因素。先进的材料将减少自身的重量,使身体相对平衡和协调,从竞技的发展中,我们获得了类似的进化和优化过程。同时,我们将所有极少主义的观点、效率、时尚、逻辑的必要性和美展现出来,这如同构想我们的建筑的过程。

建筑没有建筑师

绝大多数过去或现在的建筑并没有建筑师的规划,文化、经济、民俗、历史等原因造就了结果。作为建筑师,我们面对过去可以无能为力,而面对未来我们应该坚持原则。

职业的责任可以使建筑师坚守信念,如果每一个建筑师能够坚持原则,错误将极少发生。我们必须积累经验,尝试开辟未来建筑的途径,经验会在这具有近似意义的进化中被精简、重组、完善,最终达到科学的结果。

放松

放松是物理和心理紧张状态得以平衡的结果,我们现代人的生活状态是非常紧

EVOLUTION

张、复杂的，甚至不可预见。现代人的生活和工作常常处于紧张、苦恼、无助的状态，物质需求往往超越了心理需求，它们之间需要协调以使人放松。

建筑可以减少压力，平衡心理，使我们的生活状态放松，使生活趋于简单和轻松。人们重视建筑的原因在于它是我们生活的基础，它是构成城市最基本的元素，它在一种平衡的状态和环境下不断发展着。

转换

"Entropiee"在希腊语中是"转换重归"的意思。转换在物理学中是一个热动力和不可逆转的量的系统。在一个封闭的系统里，在条件允许的状态下产生转换，周围的每一个事物将发展为平衡的状态，平衡是转换的结果。

极少主义

建筑在非紧急的状态下将外表形象和体量弱化，以达到与周围环境的和谐，称为"极少主义"。这里的极少主义的概念不仅限于形式，而是在能源方面、经济方面也要遵循极少主义原则。

2. 滑雪
3. 香港九龙住宅
4. 西双版纳的傣族民居，这一传统的木结构住宅已有一千多年的历史
5. 蜂巢

2. Skiing
3. Residential area in Kowloon, Hong Kong – 1000 years without architectural planning
4. Xishuangbanna Dai houses, the traditional wood frame house than a thousand years of history
5. Honeycomb

1 进化
EVOLUTION

6
6. 新北京四合院
7. 根据出土文献绘制的复原图，乌克兰，年代约公元前28 000年。场地尺寸：12m×4m。图片：DETAIL Zeitschrift fuer Architetur Serie 2000, 6
8. 公元前700年的尼尼微石雕
9. 蒙古包是蒙古族的住房，它的图案对称、美观，结构简洁，在游牧时可以随拆随立，适应游牧生活，图片为一家人搭建的蒙古包
10. 奥迪场馆——进化后的结果

6. New Beijing courtyard
7. Reconstruction of tent based on remains found in Ukraine, ca.28,000 BC, plan dimensions:12x4m. From: Andrew Sherratt(eg). Picture: DETAIL Zeitschrift fuer Architetur Serie 2000, 6
8. Tent in Nineveh, ca, 700 BC.
9. Yurt is a Mongolian housing, symmetrical pattern it beautiful, simple structure, with the legislation with the demolition of the nomadic is to adapt to the nomadic life, the picture for a person to build yurts
10. Audi venues - evolutionary results

Evolution, derived from the Latin "evolvere", refers to a process of gradual and progressive change (over millions of years) in the gene pool of a population by environmental alteration, natural selection and genetic drift. All life forms experience evolution.

The process of evolution
In the model of scientific experiments, we simplify the evolutionary process as a continuous chain or a spiraling disk component. Due to outside influence, the evolutionary process starts with the change of the stable status into the unstable, and then triggering a new structural generation, followed by the demise of the old structure and the formation of a new status which becomes stable. This status further develops and blends into the new circulation system. Therefore, it can be inferred that the evolutionary system updates its internal organization while seeking the stability in the controversial status. In principle, a large amount of data indicates the existence of the internal competing yet simultaneously mutual satisfying conditions within the evolutionary status.

Strength
energy, technique, concision and nature, these are the essences skiing embraces. The application of new technologies from various fields has brought revolutionary development to this athletic sport: lighter ski equipment and skiwear improves the balance and speed in jumping and sliding, and enhances the experience of thrill, elegance, freedom and dynamic. Skiing is a manifest of evolution and optimization. The ultimately achieve scientific results.
The vast majority of the past or present buildings are created without the careful planning by architects due to economic, cultural, or historic reasons and folklore customs. As architects, we cannot change the past but the future by adhering to the principles.
Professional responsibilities allow architects to keep their faith. If every architect were able to adhere to the principles, errors could be minimized. We need to accumulate experience and to try to open up the way of the future architecture. Mimicking the evolutionary process, the experience will be streamlined, restructured and improved,

7

ultimately reaching to the scientific result.

Relaxation

The concept of relaxation is to lose the physical and psychological tension and to regain the balance. The modern lifestyle is very tense and complicated, even unforeseeable. People suffer under the tension between life and work – the state of anguish and helplessness is common – and their physical requirements often exceed the psychological, thus the adjustment between them is needed to build the basis of the relaxation.

Architecture can help relieving the pressure to achieve psychological balance, and therefore leading our life into the state of relaxation – life tends to be simple and easy. People pay attention to the architecture because it is the foundation of our life, constituting the basic elements of the city, and it continuously develops in a state of balance.

Transform

The conversion concept derived from the ancient Greek entropie "conversion reintegration" mean.

Conversion system of a thermal power and irreversible amount in physics. In a closed system, in the state of conditions allowed the conversion around every thing will be developed for the balance of the state, the balance is the result of the conversion

Minimalism

Minimalism in architecture weakens the external image and the body mass in unforced condition to reach the harmony with the surrounding. The generalized touch of minimalism is wide and broad. It is the awakening of the living attitude, the responsibility to energy consumption, the respect and awe paid to nature, and the appreciation and cherishing of life. Hence our architecture should answer the spirit of the time and highlight the charm in harmonic symbiosis with nature.

8

9

10

项目解读:重庆自然博物馆新馆
建 筑 师:卢卡斯·齐姆尼
CASE STUDY: NATURAL HISTORY MUSEUM, CHONGQING
ARCHITECTS: LUKAS ZIMNY

NATURAL HISTORY MUSEUM, CHONGQING

重庆自然博物馆新馆工程项目选址于素有"嘉陵江畔明珠""重庆市后花园"美誉的著名风景旅游区——重庆北碚。具体位置在重庆市北碚新城。项目地点面靠缙云大道，背倚缙云山，风景优美，交通便利。

博物馆整体造型尊重城市规划，与周边环境相协调。博物馆主体转折顺应山势，形成广场，集散入口人流、车流，自然划分了室外展场。各区域交通流线清晰，互不干扰。南向地势开阔，临近交通干道和公共交通设施。结合名人雕像园，设计主入口广场。入口左侧结合地形设计阶梯状景观，布置植物观赏区及儿童乐园。结合基地东南侧山形，布置动物观赏区和恐龙生态园，以自然地形遮挡动物观察区所必需的围栏，不影响建筑整体外观。在基地北侧制高点设计观景平台，俯视可总览博物馆全局，眺望可观赏缙云山景色。基地西侧为车行主入口，布置室外停车场，并设计有大型客车停放区域。东侧为工作人员出入口和货运出入口，可直达办公科研区及藏品库。精心设计总平面竖向，所有坡度顺应地势向基地西南角降低，室外地坪排水顺畅。

博物馆各层展览功能及相应配套设施完善、独立，可分期实施而不影响整体形象。展览陈列、科研办公、藏品收藏各功能区分区明确，流线互不交叉，易于管理。主入口大厅位于二层可直达各公共服务用房和报告厅。通过连桥，穿越人工溪谷，可到达各陈列区及影院。

临时展厅位于一层，拥有独立出入口，也可从主入口大厅到达。博物馆参观流线，序厅－"地球厅"－"恐龙厅"－"西部厅"－二层环境厅－进化厅－生物展厅－山水展厅。展览运用声、光、电、多媒体虚拟技术及互动参与等多种展览手段，构建主题鲜明、内涵丰富的展览空间。各厅均与自然环境结合，绿化从两侧层层渗入，室内与室外、环境与建筑完美交融，参观者可在建筑中体验自然。

休息中庭连接各陈列厅，既是交通枢纽，也是休憩停留、观赏体验自然环境的核心场所。中庭周边为全落地玻璃幕墙，开敞通透，可直达人工溪谷。建筑间连桥穿越溪谷上空，更进一步提升了参观者的自然体验。科研区位于建筑左侧，拥有独立出入口。卸货区位于左端，货运便捷。科研用房与藏品库联系紧密，方便科研及修复人员对藏品进行管理。办公用房独立，可直达展览、科研及藏品各区，统一管理，并有独立花园阳台，办公环境舒适。藏品库开间尺寸符合标准库房要求，既可灵活布置设备，也可适应不同种类藏品的库房使用要求。藏品库通过专用货运通道和货梯，可直达各陈列室，不与公众人流交叉。

NATURAL HISTORY MUSEUM, CHONGQING

The new site of Chongqing Natural History Museum is located at Beibei, a famous scenic area reputed as the Pearl of Jialing Riverside and the Back Garden of Chongqing. Facing Jinyun Boulevard and backing Mount Jinyun, the main structure conforms to the geological setting and embraces the plaza which distributes the entrance's traffic and naturally divides the outdoor exhibition areas. The traffic flows are smooth and clear of interference.

The south terrain is open and wide, close to the main roads and public transportation, therefore designed as the main entrance plaza with a celebrity statue garden. On the left side of the entrance, the orographic landscaping steps provide botanical appreciation as well as the function of a Children's Playground. The southeast hill country is aligned with an animal watching zoo and a dinosaur biological garden. The natural topography screens the zoo from the main architecture appearance, contributing one of the major design highlights. To the north, the commanding height turns into a scene viewing platform, overlooking the museum compound and the scenic mount Jinyun afar. The main vehicle entrance locates to the west of the terrain, facilitated with an outdoor parking lot as well as a designated zone for tourism buses. To the east of the terrain is the entrance for staff and cargo, with direct access to the research office and the collection warehouse.

The exhibition function and the supporting facilities of each floor are complete and independent, so that the exhibitions can be set up in phase without affecting the overall image. The exhibition display, research & administration and collection storage are partitioned in function for easy management. The main entrance hall is located at the second floor, directly accessing all public services and lecture halls. Across the bridge over the artificial stream valley are the display area and the movie theater. The temporary exhibition hall, located on the ground level, is accessible either through a separate doorway or from the main entrance hall.

The tour route starts from the Preface Hall, followed by the Earth, the Dinosaur and the Western Hall. Up to the second floor is the Environment Hall, then through the resting atrium with the top view of the dinosaur fossils, the route leads to the Evolution Hall, the Biological Showroom and the Landscape Hall. Exhibitions use the interactive participation of sound, light, electricity and the multimedia virtual technology to build a distinctive theme with rich connotation and generous exhibition space.

All the halls are combined with the natural environment; mountains and green layers penetrate from both sides. Indoor and outdoor, the perfect blend of nature and architecture, visitors linger here to experience Nature in Architecture or enjoy the architecture in the nature. The atrium connects each showroom, servicing as the traffic hub as well as the core resting area to stay and enjoy the natural setting. In the atrium lounge, the full floor-to-ceiling glass façade contributes to the openness and transparency, and the artificial creek valley can be accessed directly from here. The bridge over the valley further enhances the visitors' experience of nature. The research space has a separate entrance at the left side of the building. It is closely connected to the collection warehouse, to facilitate the research and restoration of collections. The unloading zone on the left is very convenient for cargo shipping. The office space with a separate garden balcony has direct access to all function zones, comfortable and supporting centralized management. The collection warehouse has the right dimension to accommodate all specimens. It can also be flexibly arranged and adapted to different storages.

总平面图 比例：1：4000   SITE PLAN SCALE 1：4000

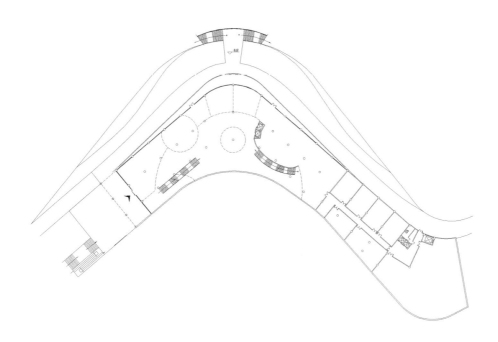

一层平面图 比例：1：1500　　GROUND FLOOR SCALE 1：1500

项目解读：贵阳医学院图书馆
建 筑 师：德雅视界建筑师事务所
CASE STUDY: GUIYANG MEDICAL INSTITUTION LIBRARY, GUIYANG
ARCHITECTS: IDRAL ARCHITECTS

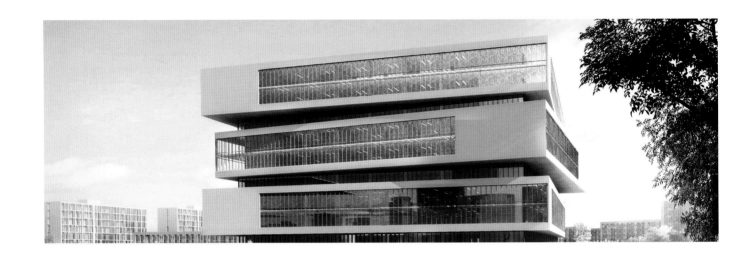

GUIYANG MEDICAL INSTITUTION LIBRARY, GUIYANG

贵阳医学院图书馆被规划为学院的标志性建筑，它的造型使观者非常直观地感受和理解。

图书馆位于贵阳新区贵阳医学院南北轴的交接点，并成为学院的中心。图书馆的造型来源于2010年上海世博会德国馆书籍展厅内书籍的叠放，它成为图书馆造型设计的原型。它简单的形体让人们直接感受内容与外形的完美统一。

图书馆平面呈长方形，共八层，北侧为主要出入口，南侧为次出入口。南北两个出入口连接了教学区和生活区，方便两个区域的学生进入图书馆。整个建筑内部设计了T形的中庭，中庭空间自六层起扩大了中庭的面积，形成阶梯状。一层中心区设计了咖啡休息区和阅览区，西侧为开放式图书阅览区，东侧是行政管理区。二层至五层为专业图书馆，同时设计了阅览区。六层至八层为数字图书馆和信息中心。

图书馆的外立面设计了玻璃和石材对比的形式，从而抽象了书籍的概念。玻璃采用的是印花玻璃，印花玻璃的图案选择了贵州蜡染的风格图样，以展示贵州的传统文化。

Planned as the centre piece of the medical institution in Guiyang, the formation architectural expression and content of the architecture are very straightforward and outspoken.

Located at the intersection of the north-south and east-west axes, the library is the center of the new Guiyang University campus. The design concept is borrowed from the idea of the stacked books of the German Pavilion at Expo Shanghai in 2010. The simple shape of the building acts as a perfect representation of the content.

The floor plan of the library is a rectangle in outline, with eight levels in total. The major entrance to the north and the secondary entrance to the south connect the teaching and living areas of the campus, providing convenient access for the students from both directions. The atrium runs through the whole building and widens at the 6th level to form a T-shaped open space. There are a cafeteria and a reading area at the center of the 1st floor, an open reading area in on the west, and the administration sector on the east side.

The façade of the library uses the contrast between glass and stone, in order to bring to mind the concept of books. Noteworthy is the printed façade glass, as the printed design is chosen from patterns of the Batik style, representing the local traditional art and culture.

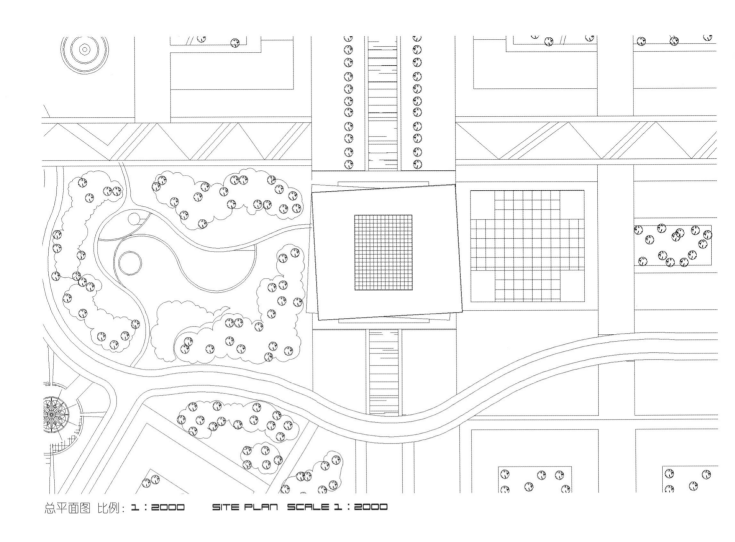

总平面图 比例：1：2000　　SITE PLAN SCALE 1：2000

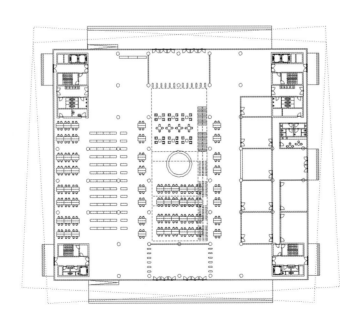

一层平面图 比例：1：1000　　GROUND FLOOR SCALE 1：1000

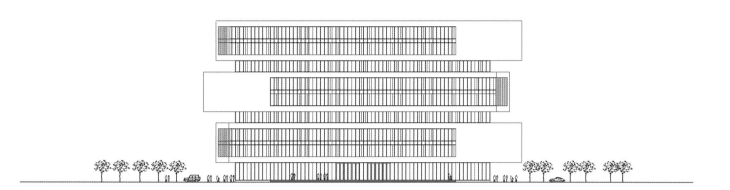

立面图 比例：1：1000   ELEVATION SCALE 1：1000

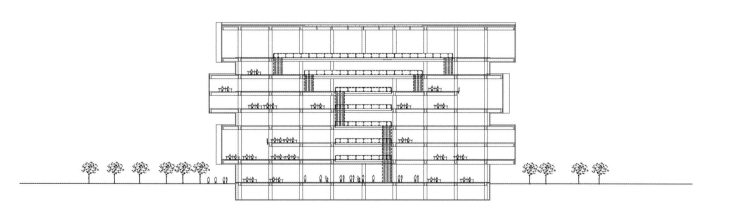

剖面图 比例：1：1000   SECTION SCALE 1：1000

2 生态
ECOLOGY

技术和生态之间没有冲突，生态理论在建筑中的应用被称为"完整系统的整体规则"，对现代科学技术的认识、判断、应用应积极考虑人文因素和人的需求。作为建筑师，我们可以将生态学的成绩转换为结果，同人、建筑、城市融合并有机利用。

THERE HAS NEVER BEEN ANY DE FACTO CONFLICT ARISING FROM BETWEEN TECHNOLOGY AND ECOLOGY. INTERESTINGLY, THE APPLICATION OF ECOLOGIC THEORIES TO THE SCIENCE OF ARCHITECTURE IS APPRECIATED AS THE GENERAL PLANNING FOR COMPLETE SYSTEM. NOWADAYS, CULTURE AS WELL AS HUMAN DEMAND IS TO BE GIVEN MORE AND MORE ROLES TO PLAY WHILE MODERN SCIENCE AND TECHNOLOGY ARE UNDERSTOOD, JUDGED AND PUT INTO PRACTICE. ARCHITECTS ARE RIGHT IN THE POSITION TO TRANSLATE THE OUTPUTS OF ECOLOGIC RESEARCH INTO PRACTICAL USE, STITCHING TOGETHER HUMAN, BUILDING AND CITY.

2 生态
ECOLOGY

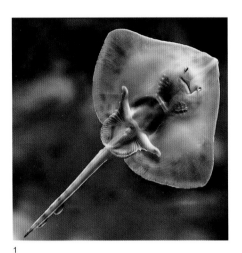

1. 魟鱼是一种重要的软骨鱼，其品种有500多种

1. Batoidea is a superorder of cartilaginous fish commonly known as rays and skates, containing more than 500 described species in thirteen families

生态

"生态"一词来源于古希腊语，生态的概念是指家（house）或者我们的环境。

生态系统是由生物群落和它的无机环境相互作用而形成的统一整体，是在一定时间、空间内由生物群落与环境组成的一个整体，各组成要素间借助物种流动、能量流动、物质循环、信息传递和价值流动而相互联系、相互制约，并形成具有自我调节功能的复合体。

生态学

生态学是研究有机体及其周围环境相互关系的科学。

随着人类活动范围的扩大与多样化，人类与环境的关系问题越来越突出。因此近代生态学研究的范围，除生物个体、种群和生物群落外，已扩大到包括人类社会在内的多种类型生态系统的复合系统。人类面临的人口、资源、环境等几大问题都是生态学的研究内容。

人与自然

生态学是一个源自于生物学的学科，它研究的内容包括：物质、生活环境、活体和死体的转换关系。同时它研究时间的发展和转折、争议及如何重新塑造现今平衡的生态系统。

生态学展示了人、环境、技术综合的转换关系，研究现今人类的综合问题，并提出解决方案，它服务于多种学科、支持多项研究。

生态建筑

所谓生态建筑，是根据当地的自然生态环境，运用生态学、建筑科学技术的基本原理和现代科学技术手段等，合理安排并组织建筑与其他相关因素之间的关系，使建筑与环境成为一个有机的结合体，同时具有良好的室内气候条件和较强的生物气候调节能力，以满足人们居住的舒适性，使人、建筑与自然生态环境之间形成一个良性循环系统。生态建筑其实是"更高级的适者生存建筑"。

生态学在建筑中的应用

我们人类的生态环境在近200年经历了前所未有的巨大变化和破坏，而人为因素，如城市化、工业污染、战争均是加快生态环境突变的元凶。二氧化碳的排放加快了气候的变暖，导致了众多自然灾害的发生。保护我们赖以生存的生态环境是地

2

球人不可推卸的责任。

建筑作为承载个体的空间，是否以健康、生态的形式设计，将直接影响到人们的生活质量、生活节奏及身体健康。所以，综合利用现代节能技术和节能产品、保护自然是建筑师义不容辞的责任。

保护自然资源、使用再生资源和新能源，如太阳能、风能、地热及其他能源系统等，是新一代建筑师的责任和任务，同时建筑师必须综合利用现代节能技术和节能产品。

重视大楼使用者的综合感受，如日光、自然空气的质量、室内温度与湿度、色彩的整体质量，并通过技术手段来实现。

最终，我们希望建造一个与自然生态系统和技术系统融合的现代建筑，使使用者能够长期感受到建筑的美与舒适。

3

4

5

2.西藏的生态建筑 ——林谢德太阳能学校 图片：DETAIL Zeitschrift fuer Architetur,Serie 2002, 6
3.生态建筑 ——德国杜塞尔多夫城市水电供应公司生态中庭 建筑师：英恩霍文建筑师事务所
4.生态建筑 ——新加坡罗宾逊路71号 建筑师：英恩霍文建筑师事务所
5.生态建筑 ——成都恒泽动力办公楼 建筑师：德雅视界建筑师事务所

2.Lingshed Solar school, Tibet. DETAIL Zeitschrift fuer Architetur Serie 2002, 6
3.Eco-Building-Stadtwerke, Duesseldorf Germany. Architects:Ingenhoven Architects
4.Eco-Building-71.Robinson Road, Singappre. Architects: Ingenhoven Architekts
5.Eco-Building-Hengze Building, Chengdu. Architects: ideal Architects

2 生态
ECOLOGY

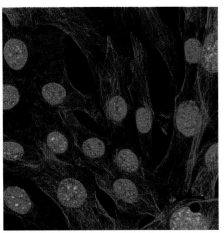

6
6. 这是细胞（纤维原细胞）的图像，其中细胞膜上好像有多处穿透孔，细胞的组成部用带颜色的抗体对照或使用特殊的颜色着色，这是上下重叠的图像
7. 生态建筑——新加坡玛丽娜湾1号 建筑师：英恩霍文建筑师事务所
8. 生态建筑——悉尼布莱街1号——双层玻璃幕墙系统是通过中庭进行自然通风 建筑师：英恩霍文建筑师事务所

6. Cell image (here is a fiber blast cell) It looks like many penetrating holes in the cell membrane, and the cell component is contrasted by the colored antibody or tinted with special color. This is the vertically overlapped image
7. Eco-Building-Marina One, Singapore. Architects: Ingenhoven Architects
8. Eco-Building-1 Bligu Street, Sydney. Its fullyglazed tower is equipped with a double-skin façade and ventilated by an atrium that rises the building's entire height. Architects: Ingenhoven Architects

Eco- (Okios)
Derived from the ancient Greek, the concept of "ecology" refers to a household or our environment.
An ecosystem is a community of living organisms in conjunction with the nonliving components of their environment interacting as a system in limited time and space. These components are linked and constrained through species flow, energy flux, nutrient cycles, and information exchange, forming a self-regulated complex.

Ecology
Ecology is the scientific study of the relationships that living organisms have with each other and with their natural environment.
With the expansion and diversification of human activities, the relationship between human beings and the environment is becoming more and more prominent. Therefore, the scope of modern ecological studies, in addition to individual organisms, populations and biomes, has extended to the complex system of various ecosystems including the human society. Humanity concerns such as population, resource and environment, are all studied in Ecology.

Man and Nature
Ecology is a subject originated from biology; its researches incorporate the transforming relationship between substances, living environments, as well as live and dead bodies. At the same time, it studies the development of time and its turning, the current situation of questions, and how to reshape the present balance of the ecosystem.
Ecology reveals the integrated transforming relation between people, environment and technology; it studies today's human problems and proposes solutions; it serves various subjects and supports a number of studies.

Eco-Building
Eco-building refers to a structure that acclimatize its local natural ecological environment, applies ecology and architecture principles and modern science and technology, coordinates other relevant factor, so that the structure is organically

embedded with its surrounding, enabling a stronger bioclimatic ability to regulate the indoor climate to meet the resident's living comfort, therefore human, architecture and natural ecological environment form a virtuous cycle system.Indeed, eco-building is a more advanced architecture of "survival of the fittest".

Ecology in Architecture

Our human ecological environment has experienced unprecedentedly tremendous changes and violations in the last 200 years. Human factors, such as urbanization, industrial pollution and war, are all culprits accelerating the mutation of our ecological environment. Carbon dioxide emissions accelerate the global warming, leading to the occurrence of many natural disasters. It is everyone's inevitable responsibility to protect the ecological environment that our survival relied upon.

As the bearer of individual space, whether the architecture is designed in form of a healthy ecology will have direct impact on the quality and pace of our lives as well as our health condition. Therefore, it is the bounden duty of an architect to comprehensively utilize modern energy-saving technologies to protect the nature.

It is our responsibilities and tasks as architects of the new generation to protect the natural resources, to use renewable resources and new energies, such as: solar, wind, geothermal energy etc. and to utilize comprehensively the modern energy-saving technologies as well as the energy-saving products.

We value the user's experiences and feelings towards the building, such as: sunlight, natural air quality, indoor temperature, humidity, the work environment, the overall quality of the color. And we try to achieve the goals through technical means.

Ultimately, we hope to construct a modern building fusing the natural ecological systems and the technical systems, so that its user could experience the beauty and comfort of architecture in the long run.

9. 生态建筑 ——苏黎世施华洛世奇办公楼，暖气和冷气源于苏黎世湖水　建筑师：英恩霍文建筑师事务所
10.生态建筑 ——海口逸龙广场　建筑师：德雅视界建筑师事务所

9. Daniel Swarovski Corporation, Zuerich Water from the lake is used both for heating and cooling. Architects: Ingenhoven Achitects
10. Eco-Building-Yulong Plaza, Haikou. Architecets: Ideal Architects

项目解读: 德国莱茵集团总部大楼
建 筑 师: 英恩霍文建筑师事务所
CASE STUDY: HEADQUARTER BUILDING OF RWE, ESSEN, GERMANY
ARCHITECTS: INGENHOVEN ARCHITECTS

HEADQUARTER BUILDING OF RWE, ESSEN, GERMANY

大楼位于一个点状的低层建筑群中，周围有小型的湖泊和绿色的草地，以节能为前提的建筑平面呈圆形，而圆柱形的形体易于减轻风力和热能的损失。

建筑双层玻璃的设计使内部空间进行自然通风，因为外墙的玻璃已经阻挡了高空的压力，而位于30层的屋顶花园的玻璃外墙则起到了阻挡风力的作用。

双层玻璃幕墙是由外层玻璃、内层玻璃及通风构件组成的，位于内外层之间的空间，提供了一个气候保护的可能。夏天阻挡外部的热量，冬天阻挡寒冷并起到通风的作用。外层玻璃的厚度是10mm，大小是2m×2.6m的单面安全玻璃，每块玻璃由八个点状构件固定，内层玻璃是与楼层同高的推拉窗，它可以打开到135mm，空气走廊宽度为50cm。

鱼嘴构件起着与空气交换的作用，其形状有利于大量的光线进入室内。新鲜的空气流进鱼嘴构件内停留后进入空气走廊，然后再进入室内。同样，室内的空气是通过空气走廊进入鱼嘴构件的，然后排出，形成自然的通风系统。

The RWE headquarter building sits among a group of punctate low building with a small lake and green grass land. The circular design is superior on the premise of energy saving in architectural shape and equipment; and the cylindrical modeling mitigates the wind impact to reduce the heat loss.

The double glass facade is made of two layers of glass and the ventilation components. The outer layer of the double glass facade blocks the wind pressure in high air, allowing natural ventilation by opening the inner glass window. The roof garden on the 30th floor is shielded by the double glass facade as well. The air gallery between the outer and inner layer enables climate control, ventilating while blocking the summer heat and winter chill. The outer layer uses 10mm unidirectional safety glass, 2m x 2.6m by piece and anchored by eight point-joints. The inner layer is constructed by push-pull windows aligned with the height of each level, open up to 135mm. The air gallery is 50mm wide.

The fish mouth components serves air circulation and facilitates natural lighting. Fresh air flows into the fish mouth component, shortly before it reaches indoor through the air gallery. Vice versa the air indoor is discharged through the gallery then the fish mouth component, completing a natural ventilation cycle.

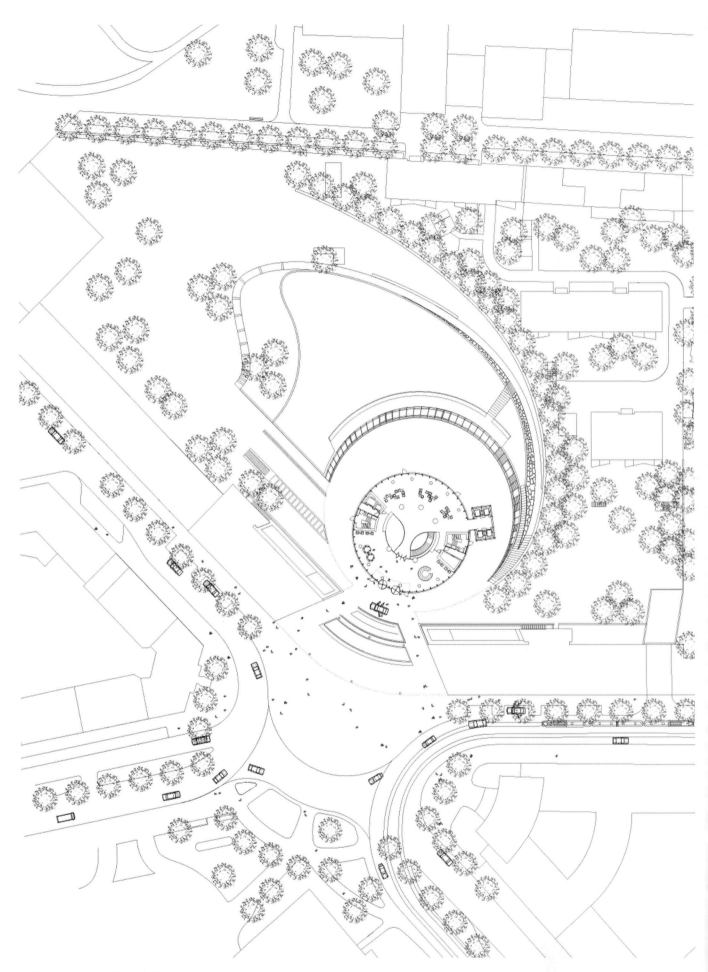

总平面图+一层平面图 比例：1：1000　　SITE PLAN + GROUND FLOOR SCALE 1:1000

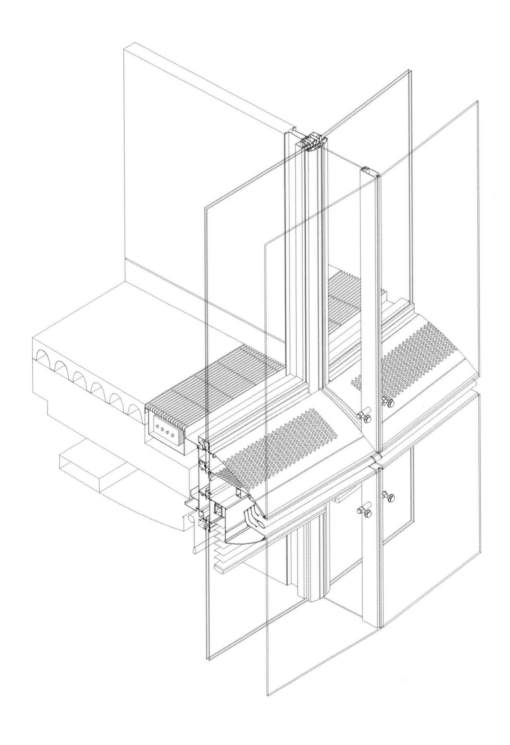

双层玻璃幕墙鱼嘴通风构件轴测图　　DOUBLE FAÇADE FISH MOUTH VENTILATION COMPONENTS ISOMETIC

双层玻璃幕墙节点模型　　DOUBLE FAÇADE FISH MOUTH VENTILATION COMPONENTS MODEL

3 能源
ENERGY

越来越多的人在城市里居住和生活，大量的资源和能源消耗给人类带来了生存危机，重视资源、保护环境是我们每个人的义务和责任。一个优秀的建筑应该能够均衡能源，将自己的能源需求通过使用新能源，如太阳能、地热，达到循环的状态，使能源消耗和能源再生保持平衡。

未来建筑可以成为一个收集器，其自身可以发挥收集、生产、使用的功效。

MORE AND MORE PEOPLE COME TO SETTLE DOWN IN CITIES, WHICH GIVES SIGNIFICANT RISE TO THE CONSUMPTION OF RESOURCE AND ENERGY AND PUTS THE SUSTAINABILITY OF HUMANITY IN GREAT JEOPARDY. IT FALLS ON EVERY HUMAN INDIVIDUAL ON THIS PLANET TO TAKE UP THE RESPONSIBILITY TO PRESERVE RESOURCES AND PROTECT THE ENVIRONMENT.
FUTURE ARCHITECUTE WILL BE A CONTAINER, WCHICH COULD COLLECT, PRODUCE, AND BE USED.

3 能源
ENERGY

1. 北京昌平低耗能别墅酒店　建筑师：德雅视界建筑师事务所
2. 欧洲投资银行总部大楼，新时代节能建筑　建筑师：英恩霍文建筑师事务所
3. 新能源——风能
4. 新能源——太阳能
5. 新能源——地热

1. Beijing Changping low energy house. Architects: Ideal Architects
2. New era of energy-efficient buildings Enropean Investment Bank. Architects: Ingenhoven Architects
3. New Energy - Wind energy
4. New Energy - Solar energy
5. New Energy - Terrestrial heat

"能源"一词来源于古希腊语"energeia"，是"影响的力量"的意思。能源是可以直接或经转换提供人类所需的光、热、动力等任一种形式能量的载能体资源。确切而简单地说，能源是自然界中能为人类提供某种形式能量的物质资源。世界能源委员会推荐的能源类型分为：固体燃料、液体燃料、气体燃料、水能、电能、太阳能、生物质能、风能、核能、海洋能和地热能。

地热

我们以地热为例，据权威机构估计，地热的总蕴含量约为地球煤炭总能量的1.7亿倍，除了人们所熟知的温泉外，地热能源更有取暖、发电等功效。目前世界上少数国家研究的增强型地热系统，有望在不远的将来替代人类的基础能源。

在德国杜塞尔多夫的住宅项目里，建筑师构思了一个节能方案。通过热泵、地热设备获取热能，用热能探针抽取地下水的热能（热交换），然后通过热泵将热能引入建筑的散热设备，建筑的散热设备被安装在中轴墙内，由它将热能传递到每一个房间，同时在外墙的窗台下也安装了同样的散热器来阻挡来自玻璃的冷气，地下取热方式不仅解决了建筑的取暖问题，也降低了地下水的温度，同时降低了地球的温度，是一个积极的生态节能方法。

可再生能源

可再生能源是指在自然界中可以不断再生、永续利用、取之不尽、用之不竭的能源，它对环境无害或危害极小，而且资源分布广泛，适宜就地开发利用，主要包括太阳能、风能、水能、生物质能、地热能和海洋能等。

能源供给已经不再是简单的供电、供水、排污问题，它已经成为了一个生态平衡的系统问题。未来能源的供给方案应该能达到自然的循环状态。德国Hammfeld的新城区规划中，建筑师提出了一个未来型的能源使用构想，在使用再生能源（风能、太阳能）的基础上，还集合了垃圾处理系统来支持这一地区的部分能源需求。具体做法是：在这一区域内除了采用热能、风力发电，还将生活垃圾经过加工处理，转换为生态燃气或作为肥料用于植物，它们同淤泥释放气体一同转化为其他能源形式。

The concept of energy evolved from the ancient Greek "energeia", converted to meet mankind needs, in any form such as light, heat and power. To be simple yet accurate, energy is the resource in the nature that provides mankind power in some form. Recommended by the World Energy Council, the types of energy are listed as solid fuels, liquid fuels, gaseous fuels, hydropower, electrical

ENERGY

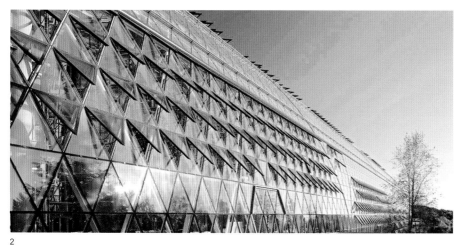

power, solar energy, biomass, nuclear energy, ocean energy and geothermal energy.

Geothermal Energy

Geothermal energy is estimated 170 million times the volume of coal energy on the earth. Geothermal is popularly linked to geysers, but it can used in heating and power generation as well. Currently, only few countries are doing research on enhanced geothermal systems, but hopefully the research would lead to an alternative solution to mankind's basic energy demands in the near future.

In a residential project in Dusseldorf, Germany, an energy-saving plan is conceived to collect thermal energy using heat pump and geothermal equipment. First, the geothermal probe is used to siphon the thermal energy (thermal exchange), then the heat pump conduct the heat into the building's radiation system. Radiators are buried within the axial wall, distributing heat to every room. They are also installed under the sill of facade windows, warming up the cold air near window glass.

Renewable Energy

Renewable energy is derived from natural resources that are replenished constantly, such as sunlight, wind, water, biomass, geothermal heat, ocean tides and waves. Renewable energy has minimal impact on environment since it is naturally regenerated and virtually inexhaustible. Geographically it is widely distributed, suitable to deploy locally.

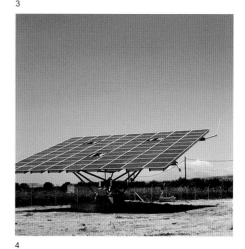

Energy supply no longer simply deals with electricity, water and sewage. It has to face the systematic challenge of ecological balance. Future energy supply solutions ought to achieve the balance of natural circulation. In the new city planning of Hammfeld, Germany, architects drew a blue print of future energy usage: combining renewable energy with waste disposal to support local energy supply. Besides solar and wind power, everyday waste disposal is processed and converted into fertilizer or ecological gas which can be further utilized in various forms.

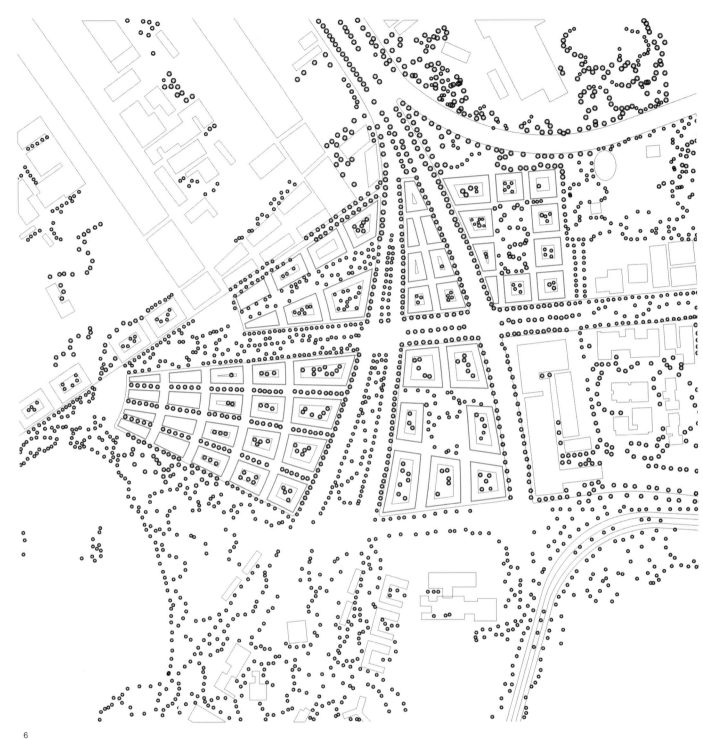

6

6. 德国Hammfeld的新城规划　建筑师：英恩霍文建筑师事务所
7. 新北京四合院，根据德国被动式节能建筑规范设计　建筑师：德雅视界建筑师事务所
8. 太阳能光电接收板
9. 能源循环示意图
10. 德国杜塞尔多夫节能住宅　建筑师：英恩霍文建筑师事务所
    德国杜塞尔多夫节能住宅图例
    1-地下水源
    2-热交换
    3-热泵
    4-散热墙
    5-加热外墙
    6-通风区域

6. New City Planning, Hammfeld, Germany. Architects: Ingenhoven Architects
7. Beijing Courtyard, Specification designed according to passive house in Germany. Architects: Ideal Architects
8. Solar Schematic diagram
9. Energy Cycle diagram
10. Energy-efficient house.Duesseldorf, Germany. Architects: Ingenhoven Architects
    Energy-efficient house legend
    1-Groundwater sources
    2-Heat exchange
    3-Heat pump
    4-Cooling wall
    5-Heated exterior wall
    6-Ventilated area

7

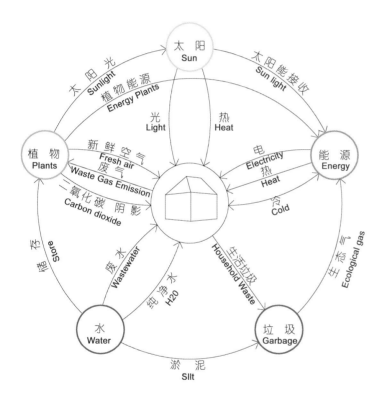
9

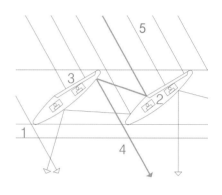

8
太阳能光电接受板图例
1-构造
2-百叶
3-接受板构件
4-日光折线
5-照射线

Photovoltaic module legend
1-post
2-louvre
3-Photovoltaic module
4-Sunlight
5-Irradiation

10

项目解读: 德国法兰克福商业银行
建 筑 师: 英恩霍文建筑师事务所
CASE STUDY: COMMERZBANK, FRANKFURT, GERMANY
ARCHITECTS: INGENHOVEN ARCHITECTS

COMMERZBANK, FRANKFURT, GERMANY

德国法兰克福被称为"银行城市",在过去的几十年里,德国商业银行(Commerzbank)就是其中之一。它的旧楼始建于20世纪70年代,法兰克福金融业的发展促使了德国商业银行新办公楼的建成。

克里斯多夫·英恩霍文(Christoph Ingenhoven)建筑师事务所构想了一个185m的高层建筑,其中众多的空间功能需求及4000多名员工的工作必须在一个相对较小的基地内完成,这显然十分困难。大楼设计了52层,一层的大堂具有四层高的空间并且留有通道通向旧楼。

基于对生态、能源的要求,建筑师团队克里斯多夫·英恩霍文协同德国著名结构工程师弗莱·奥托(Frei Otto)共同发展了这一生态建筑。弗莱·奥托长期研究材料的应用和生态建筑的发展,他对结构形式的创造是在进化及与自然环境相协调的条件下建立的。

商业银行的外形是圆柱形,通过双层玻璃幕墙的使用,可以放弃部分机械的通风系统,所有的办公空间均可获得自然的通风和自然的光线。交错排列的空中花园可以作为大楼的缓冲空间,它既提供给办公室自然的空气,又阻挡了夏天的热流和冬季的寒流。

四组三角形的柱网起到了剪力墙的作用,并且有效地组织了自由的楼层平面布局,呈交叉形的办公层部分为全层,部分为空中花园。虽然每层平面的组织各不相同,但每层办公空间均可以得到来自过道和外墙的自然光线。

建筑师在这一项目中不必强调形体在天界线和群楼中的形象,更多的是通过光线、整洁及生态的内容来表现建筑。

Frankfurt is called the "Bank City" in Germany, and the German commercial bank (Commerzbank) has been one of the contributors in the past several decades. The old building was built in the 1970s, and the new office building of the German Commerzbank rose with the financial boom in the 1990s.

To accommodate 4000 plus employees and meet various space functional requirement with much limited lot area, Christopher Ingenhoven architects envisioned a 185m, 52-story high-rise building, with a four-story space atrium in the first floor lobby, and passages leading to the old building.

Responding to the ecology and energy requirements in the tender, architects of Ingenhoven teamed up with Frei Otto, a famous German structural engineer, to work on the eco-building. Frei Otto had studied the application of materials and the development of eco-building for long time. His creation of structure adapted the theory of evolution, established in accordance with the natural environment.

The cylindrical Commerzbank with the double glass facade reduced mechanical ventilation system and supplied all office space with natural ventilation and natural light. The interleaved hanging garden was the buffer of the building, supplying natural air but leaving out summer heat and winter chill.

Four groups of triangular column grid functioned as the structural shear wall, and effectively organized the flexible floor layout- some floors are full of offices, some are part of the Hanging Garden. Among the various layouts for different floors, all office space shared natural light both from the hallway and the façade.

In this project, the architects de-emphasized the outstanding image among the skyline and the building group. Instead, they express the architecture with light, cleanliness and ecology.

总平面图 比例：1：2500　　SITE PLAN SCALE 1：2500

平面图 比例：1：1000　　PLANS SCALE 1：1000

4 城市与城市发展
CITY AND CITY DEVELOPMENT

21世纪，世界正经历着革命性的变化，虽然我们对此反应迟缓，但它是一个历史性的转折点，人们第一次经历了城市人口超越农村人口的现实。21世纪是一个真实的"城市世纪"。

The world has undergone a revolutionary change, to which we are slow in reaction. Yet it is a turning point of history - for the first time the urban population surpassed of the rural population, the 21st century is a real city century.

4 城市与城市发展
CITY AND CITY DEVELOPMENT

1. 德国法兰克福
2. Schibam(也门)——世界文化遗产城市,一个名副其实的传奇城市。这座城市由黏土城墙包围,城内有许多400年到500年历史的六层黏土高楼建筑。图片:德国国家地理
3. 杜塞尔多夫城市改造——媒介港
4. 汉堡城市改建
5. 北京
6. 可持续发展原则

1. Frankfurt, Germany
2. Schibam (Jemen) – the World Heritage city is a truly legendary city. The city, surrounded by clay walls, has many six-story high-rise buildings 400 to 500 years of history to date. Picture: National Geographic Deutschland
3. Urban renovation-Duesseldorf Medium Hafen
4. Urban renovation-Hamburg Hafen
5. Beijing
6. Sustainability Objectives - Principles

"城市"一词源于拉丁语"Urbs"。

"酷"的城市
"酷"的城市应该具有合适的规模、合理的城市规划、现代的交通系统、深厚的文化积淀、安全的城市保障和充满活力的发展空间,同时拥有认同城市文化的人才后续、可持续发展的生态城市规划蓝图和旧城保护及改造的原则与规划。我们所需要的"酷"城市是尊重历史、展望未来,具备"技术、人才、宽容和无限生机"的城市。出自时代精神的建筑构筑了城市,使城市具有了艺术的魅力。

21世纪的现代人需要的城市不仅仅是城市的历史和文化及标志性建筑,更多的是需要一个充满激情、充满机会、时尚和宽容的生态城市。

城市发展
城市发展是城市经济、社会、建设三位一体的统一的发展,包括政治、经济、社会、空间、生态、景观、文化、风尚,涉及制度、文化、科技、生态、伦理、艺术、信仰等方方面面,城市空间的规划与建设成为发展最直观的表现,和谐则是可持续城市发展的永恒的主题。

城市扩展
老城市在适合经济和社会发展的背景下,应予以改造和完善。城市的发展与扩展,应遵循生态、经济的原则,整体规划、逐步发展、限制城市过度的开发与发展,防止巨型城市的增多。

未来城市的发展
一个未来城市的发展,不再是通过大量建筑来实现的,也不是不断实现完善的城市规划,而更多的是通过政府机构、企业和公民组织共同协调来实现的。

可持续的城市发展
城市发展是一个持久性的全社会的共同任务,可持续性城市发展建立在完整的科学的、城市规划的基础上,并适合于各个层面、各个区域的经济发展,要使城市可持续发展有效,必须使社会的容量、经济、生态及文化共同发展,同时公民要承担保护环境和节约资源的责任。

CITY AND CITY DEVELOPMENT

2

人们希望在一个美丽的城市居住和生活，喜爱过去的优秀建筑，也希望更多新的优美的建筑产生。

3

4

Sustainability Objectives - Principles
可 持 续 发 展 原 则

1. Zero Carbon——零碳
2. Zero waste——零浪费
3. Sustainable Transport——可持续交通
4. Local and Sustainable Materials
   当地和可持续发展材料
5. Local and Sustainable Food
   当地和可持续发展食品
6. Sustainable Water——可持续发展水
7. Natural Habitats and Wildlife
   自然栖息地和野生动物
8. Culture and heritage——文化和文化遗产
9. Equity and Fair Trade——权益和公平贸易
10. Health and Happiness——健康及幸福

6

5

CITY AND CITY DEVELOPMENT

7

7. 丽江古城 ——世界文化遗产城市 ——始建于13世纪后期,是一个保留完好的古建筑群
8. 花园城市 ——海德堡,海德堡是在二战中仅有的两个未被盟军轰炸的城市之一
9. 麦加Haram清真寺城市广场设计 建筑师:英恩霍文建筑师事务所
10. 世界上最小的大都市 ——法兰克福
11. 完美城市的典范 ——苏黎世
12. 可持续发展目标

7. Lijiang Old Town - World Heritage City - was founded in the late 13th century, a reserved nights good ancient buildings
8. Garden City - Heidelberg, Heidelberg is only two were not in World War II Allied bombing of cities.
9. Urban renovation-Haram Moschee Mekka. Architects: Ingenhoven Architects
10. World's smallest metropolis - Frankfurt
11. Perfect city - Zuerich
12. Sustainability Objectives - Development

The concept of city is derived from Latin "Urbs".

"Cool" Cities
A "cool" city is defined by proper size, fair urban planning, modern transportation, rich cultural heritage, social security and vibrant space to develop. As well it requires generations of talents recognizing and embracing the urban culture, ecological urban planning for sustainable development, regulations and programs to protect and reform the old city. A "cool" city respects the history and outlooks the future with "technologies, talents, tolerance and infinite vitality". The city is composed of architectures from the sprint of the time, radiating the charm of art.

For the modern people of the 21st century, the city means not only history and culture, represented by iconic buildings, but more the passion, opportunity, fashion and tolerance of an eco-city.

Urban Development
Urban development is the unified development of urban economy, society and construction, including prospects in politics, economy, community, space, ecology, landscape, culture and fashion. Involving all aspects of system, culture, science, technology, ecology, ethics, art, belief, etc., the planning and constructing of urban space becomes the most intuitive representation of urban development, and harmony is the eternal theme of sustainable urban development.

The old city should be transformed and improved in the context of economic and social developments. The development and expansion of the city should follow the principles of ecology, natural resources, economic and social developments, focusing on the overall planning, step-by-step development and restriction of unlimited urbanization, to prevent the increase of mega-cities.

Future Development of the City
The development of a future city is no longer achieved by a large number of buildings neither the continuously improving urban planning, but more likely carried out by

CITY AND CITY DEVELOPMENT

8

the cooperation of government agencies and investment companies, as well as the corporate responsibilities and civic organizations.

Sustainable Urban Development

Urban development is a persistent common task of the whole society. Sustainable urban development is built on the basis of integral scientific urban planning, and is suitable for all-level or regional economies. In order to make the city development sustainable, we must develop the social, economic, ecological and cultural capacities at the same pace and all the citizens should assume the responsibilities of protecting environment and resources.

9

10

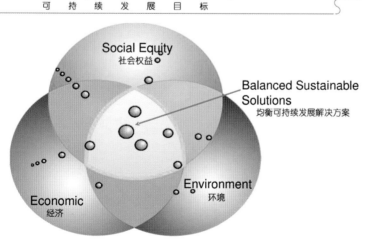
12

11

项目解读：德国杜塞尔多夫北城区CBD中心规划
建 筑 师：英恩霍文建筑师事务所
CASE STUDY: CBD DUSSELDORF, GERMANY
ARCHITECTS: INGENHOVEN ARCHITECTS

CBD DUSSELDORF, GERMANY

德国杜塞尔多夫北城区CBD中心将成为未来杜塞尔多夫市的办公中心，为世界企业及新兴产业提供了标准化、自动化、国际化的办公中心，规划题目以"Ten"(10)构成概念，意在于10分钟车程到达机场、城市中心、展览馆等重要地址，在极短的时间内联络欧洲各大城市。

CBD中心的建筑体量以高层、中层和低层构成，为世界不同企业提供了各种办公类型、层高的可能和需求。

办公机构的规划充分考虑了空间的可能性和伸缩性，租户可以随时提出自己的要求，任意搭配组合来满足国内外公司的不同需求。

绿化和水系的设计使CBD中心在花园中建立了未来办公的形式和氛围。

The CBD at the northern district is planned to be the future office center of Dusseldorf, Germany, providing standard, automated and international office space for worldwide business and emerging industries. The planning and design reflect the concept of "Ten", in which the important sites such as the airport, the expo center and downtown are all within 10 minute drive, connected to other major Europe cities in real time.

The buildings at CBD have different measures in high, medium and low rise, and the office space can be reconfigured into different size and shape to accommodate various needs for enterprises all around the world.

In perfect harmony with the landscape design, CBD creates a fine workplace environment, and activates a pleasant working atmosphere in the garden.

规划模型图片

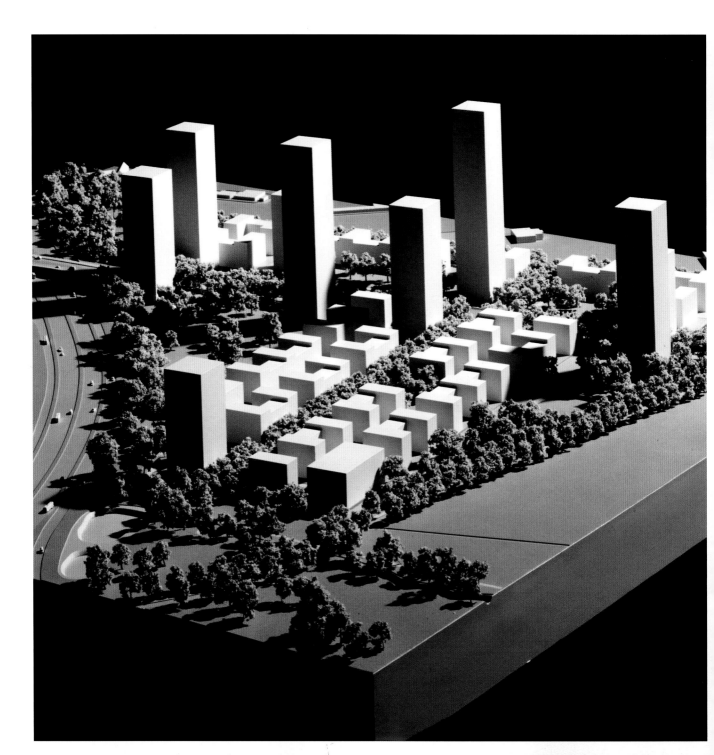

规划模型图片

5 简单与逻辑
SIMPLICITY AND LOGIC

明确、简单及易懂应该作为好建筑的前提。简单既不是新的认识又不是价值取向,它是一个希望的结果和正确的逻辑决定。

DEFINITENESS, SIMPLICITY AND LEGIBILITY SHOULD BECOME THE BASIC BENCHMARKS OF GOOD ARCHITECTURE. SIMPLICITY IS NOTHING NEWLY OBSERVED NOR DELIVERS VALUE, HOWEVER, IT IS THE OUTCOME GENERALLY EXPECTED AND DICTATED BY CORRECT LOGIC.

5 简单与逻辑
SIMPLICITY AND LOGIC

1
1. 简单的造型——天津滨海科技总部区项目
2. 模拟和数字——模拟快速直接地传递了信息，而数字传递了此时此刻精确的价值
3. 莱茵集团总部大楼模拟控制面板图
   1-灯光控制
   2-遮阳保护
   3-外墙警报
   4-温度控制
4. 莱茵集团总部大楼模拟控制面板

1. Simple shape - Tianjin Binhai Headquarters
2. Analog and digital - analog rapid and direct transfer of information, and the number passing the precise value of this moment
3. The Analog control panel, RWE headquarter building
   1- Lighting control
   2- Sun protection
   3- Façade alarm
   4- Temperature control
4. The Analog control panel, RWE headquarter building

明确、简单、易懂是好建筑的前提。它是一个希望的结果和正确的逻辑决定。

简单的建筑

简单的建筑是根据简单及逻辑的原理进行组织和设计的结果，目的是让人们直接、明确地感受建筑的外部形式和内部组织，明确自己在建筑内部的方位。简单应该作为重要的设计原则被认同。

逻辑

逻辑的概念来源于古希腊语"logos"，是"词"的意思，它的概念是易懂的说法、解释、认知、经验和使用，最初有词语、思想、概念、论点、推理之意。

模拟的大楼技术

不是每个人都能钟情于数字技术，数字技术虽然先进，但不一定完美，而模拟的大楼技术却保留了完美的可能。建筑师尝试利用模拟技术完善大楼的能源平衡、功能控制及系统组织。模拟的大楼技术应用给不同国度和地区的建筑设计质量保障提供了可能。

在莱茵集团总部大楼的项目中，建筑师建立了一个集合的控制系统，通过红外线使用面板，指令所有室内和室外的功能系统，如开窗、关窗、降下和收回遮阳帘、控制灯光、调节室内温度等，控制面板显示所有的信息和数据。这一模拟的信息系统将通过逻辑的整合最终传递给使用者。模拟的控制系统集合了现代大楼先进的机械技术和电子技术，而数字控制系统通过现代计算机程序的帮助，在模拟控制系统的基础上发展成形，发展商可以根据自己的综合状况来选择系统。

模拟和数字

技术总是偏爱精确，但人们通常依靠自己的眼睛判断尺度，其结果是大致准确。技术是数据量化的具体形式。模拟的存在和主观感受是建立在人们的客观价值中的，它不一定是适合的自然形式，相反是文化的具体选择，我们对比传统、对比系统、对比人类，或许应该在合适的时间、合适的地点及合适的条件下，模拟一个恰当的选择。

建筑的另一个概念是"恰当"，对于每一个项目和每一个难题，都有其恰当的解决方法。

SIMPLICITY AND LOGIC

2

A fine architecture has to be explicit, simple and easy to be understood. It is an expected outcome based on logic decision.

Building Simply
Building simply is to conduct design and construction based on simplicity and logic, so that the external image and the internal structure are explicitly presented, and the spatial orientation is clearly defined within the building. Simplicity should be recognized as an important design principle.

Logic
The concept of "Logic" is derived from ancient Greek "logos", originally meaning words, ideas, concepts, arguments and reasoning, but extending to notations such as statement, explanation, knowledge, experience and the usage of reasoning.

Analog Building Technology
Not everyone is attracted to digital technology. Digital technology is advanced, but not necessarily perfect. On the other hand, analog building technology reserves the possibility towards perfection. Architects have attempted to use analog technology to improve building in terms of energy balance, function control and system organization, providing the possibility of building quality control over different countries and regions.

In the project of RWE headquarter building, an integrated control system is implemented to command all indoor and outdoor function systems, such as window, curtain, lighting and indoor temperature control, via the infrared control panel. The analog control system is a collection of advanced mechanical and electronic technologies used for modern buildings, while the digital control system is developed through the help of modern computer programming. Clients and developers could choose the system based on their own comprehensive situation.

Anology and Digit
Technology always prefers accuracy, while people often reply on their own eyes to estimate the dimension, resulting in a rough accuracy. Technology is an instant

3

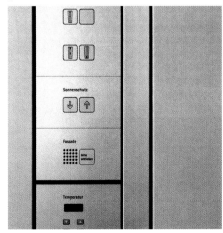

4

SIMPLICITY AND LOGIC

of data quantization. Analog presence and subjective feeling are established on one's objective value, which is not necessarily a suitable natural form but a specific choice of culture – perhaps at the right time, the right place and the right condition to simulate an appropriate choice in contrasting our traditions, in comparison of our systems and in analogy of us human beings.

Another concept in architecture is the appropriateness. For every project and every problem there is always an appropriate solution.

5

5. 简单的建筑 —— 廊坊台商产业园独栋办公楼　建筑师：中国建筑技术集团有限公司
6. 简单的形体 —— 北京普天首信集团办公楼　建筑师：中国建筑技术集团有限公司
7. 海康威视重庆基地项目　建筑师：德雅视界建筑师事务所
8. 德国奥芬堡布尔达媒介园　建筑师：英恩霍文建筑师事务所

5. Simple building - office building in the industrial park, Langfang. Architects: China Building Technique Group Co., Ltd
6. Simple shape - Beijing Putian Capitel Group office. Architects: China Building Technique Group Co., Ltd
7. The Hikvision base, Chongqing. Architects: Ideal Architects
8. Burda Medium Park, Offenburg, Germany. Architects: Ingenhoven Architects

6

7

项目解读：韩国KIM HAE体育中心
建 筑 师：英恩霍文建筑师事务所
CASE STUDY: KIM HAE STADIUM, SOUTH KOREA
ARCHITECTS: INGENHOVEN ARCHITECTS

KIM HAE STADIUM, SOUTH KOREA

KIM HAE体育中心是为2002年的韩国足球世界杯设计的主场馆，可容纳85000名观众。英恩霍文建筑师事务所被邀请参加了设计，合作的是来自英国的哈波尔德结构事务所。体育中心的要求是主场可容纳40000名观众、篮球馆容纳20000名观众附带一个训练馆、一个游泳馆、一个网球馆和一个田径馆。KIM HAE自然条件优越，体育中心的基地位于KIM HAE郊外，周围有低矮的山包，由于地理条件的限制，能够使用的平坦面积很少，建筑师必须在有限的面积内，最有效地使用空间，更重要的是要在绿色的自然环境中构筑建筑物的同时，尽可能不破坏生态环境。建筑师的目标，不是简单地在风景区内建造一幢几何形体的建筑，而是要使建筑的存在成为风景区的一部分。为与周围环境融合，建筑形式为柔美的弧形，弧形的外壳附罩在地面上，整个建筑群与大自然的绿色融为一体。该建筑群中所有建筑单体的结构形式采用简单的弧形钢结构支撑系统，屋顶的外表包括足球场、篮球馆都使用了可透光的膜结构，KIM HAE体育中心在夜晚放射出朦胧的光芒，在无垠的旷野上美轮美奂。

景观设计中尽可能地利用原有的生态环境创造自然的景观效果。KIM HAE体育中心的方案设计将建筑、风景互为背景的和谐画面展现在了世人面前，并承担起了保护生态环境的责任。

KIM HAE Sports Center is the main site for the 2002 World Cup of Soccer in Korea, accommodating an audience of 85,000 people. Ingenhoven Partners were invited to participate in the design competition, collaborating with Happold, the structure engineer firm from the UK. The sports center has a main arena accommodating 40,000 people, a basketball stadium holding 2000 with an attached training gymnasium, a swimming stadium, a tennis stadium and a track & field stadium. KIM HAE is blessed with superior natural conditions. The sports center is located at the outskirt of KIM HAE, surrounded by low hills. The accessible flat area is very limited due to the geological condition. The architects faced the challenge to make the most efficient usage of the space, and more importantly, to preserve the green natural environment as much as possible. The goal is not to simply construct any geometric shape in the scenic area, but to present the architecture as part of the scene. To blend with the surrounding the architecture is designed as a softly curved shell hooded on the ground. The entire building group is fused within the natural green. Every single construction of the buildings is supported by the curved steel structure. With the roofs of the soccer field and the basketball stadium covered by opaque membrane materials, KIM HAE sports center emits hazy light at night, emancipating ethereal beauty in the expanse of wilderness.

The landscaping design aimed to create natural effect by using the original ecological environment as much as possible. The design case of KIM HAE Sports Center presented a harmonious picture of the architecture and the beauty of nature being each other's background, and assumed the responsibility of protecting the ecological environment.

总平面图 比例：1：7500　　SITE PLAN　SCALE 1：7500

一层平面图与立面图 比例：1：3000　　GROUND FLOOR+ELEVATION　SCALE 1：3000

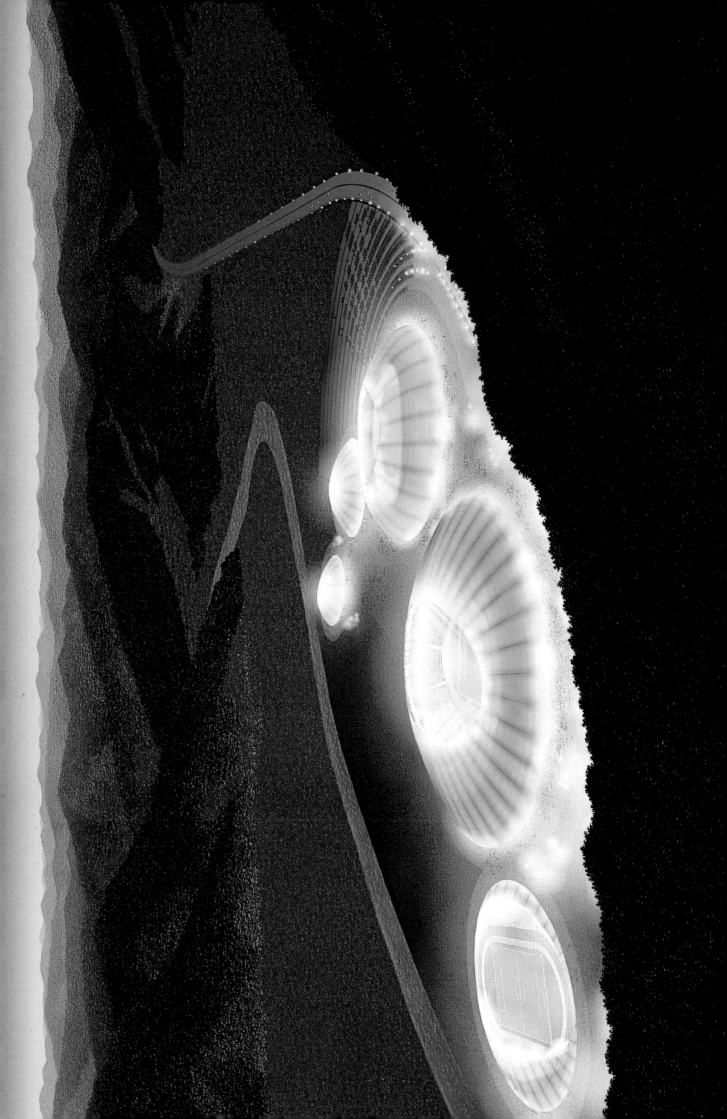

项目解读：宁波富春酒店
建 筑 师：德雅视界建筑师事务所
CASE STUDY: FUCHUN HOTEL, NINGBO
ARCHITECTS: IDEAL ARCHITECTS

FUCHUN HOTEL, NINGBO

宁波富春酒店的设计主题是在一个半圆形的基地上建立一个酒店，它将成为该地区的标志性建筑。

酒店位于宁波北仑区，基地为半圆形。由于地形的原因，酒店被塑造为弧形建筑，主楼和辅楼通过一个遮阳屋顶连接在一起。整体建筑由三部分构成：主楼19层，为酒店客房；辅楼7层，为酒店式公寓；裙房3层，为酒店裙房。

酒店大堂屋顶为全玻璃材质，表现出现代建筑的透明和时尚感。玻璃屋顶设计了金属遮阳板，遮阳板可根据太阳照射角度调节方向。大堂正立面设计了通风窗，同时玻璃屋顶设计了排风窗，形成循环的自然通风系统。

一层大堂设有传统的酒店接待、商务、商店、早餐厅及咖啡厅。二层设有各种会议室、宴会厅、健身房等酒店必备设施。三层以上北侧主楼为酒店客房，南侧为酒店式公寓。

The design theme of Fuchun Hotel in Ningbo is to build a landmark hotel a semi-circular lot.

The hotel is located in Beilun District, Ningbo, on a semi-circular lot. The outline of the building is curved accordingly, in which the main building and the auxiliary structure is connected with a sun canopy. The whole building consists of three parts: the 19-story main building hosting hotel rooms, the annexed 7-story apartments, and the 3-story hotel podium.

The hotel lobby is featured with full glass ceiling, presenting the transparency and fashion of modern architecture. The glass roof is equipped with adjustable metal visors to reflect the angle of sunlight. The ventilation widows in the front facade and the exhausters on the glass roof complete the circulation of the natural ventilation system.

The first floor lobby incorporated reception, business, store, breakfast area and café. The second floor is designed with various meeting rooms, ballrooms, fitness and other amenities. On the third floor and above, there are hotel guest rooms on the north side and apartments on the south side.

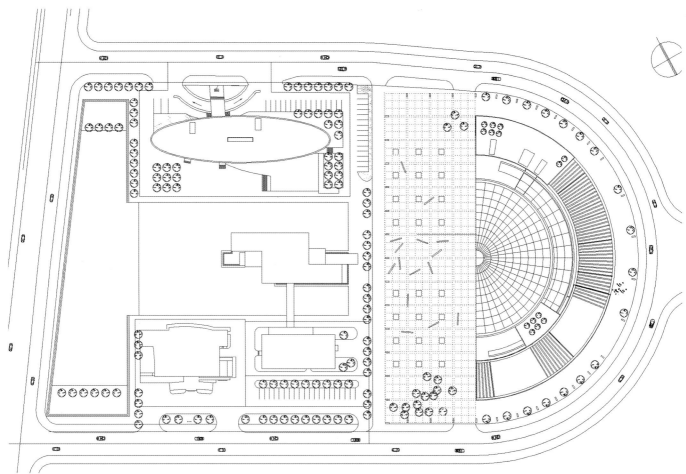

总平面图 比例：1：2000　　SITE PLAN SCALE 1：2000

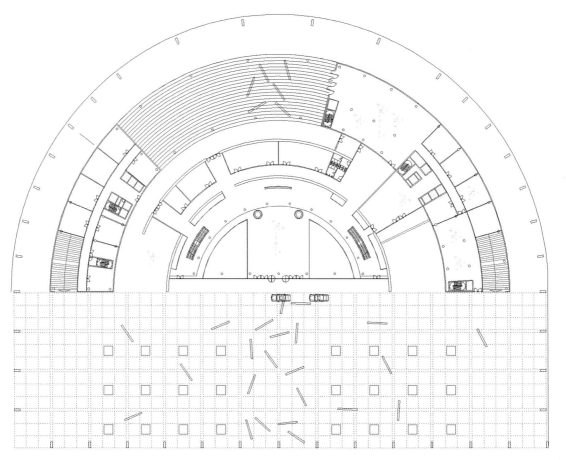

一层平面图 比例：1：1200　　GROUND FLOOR SCALE 1：1200

6 光
LIGHT

光给予我们人类希望、价值和最基本的能量。使用自然光，挖掘其价值的极限，是建筑师在生态建筑设计过程中的另一个重要课题。

LIGHT INSPIRES HOPE, DELIVERS VALUE AND FULFILS THE BASIC NEEDS FOR ENERGY. MAXIMIZING THE VALUE OF LIGHT IN NATURE HAS BECOME ANOTHER TOPIC CRITICAL IN ECOLOGICAL AND SUSTAINABLE BUILDING DESIGN FOR MODERN ARCHITECTS.

6 光
LIGHT

1
1. 通常光在380～780nm长的电磁波波长（对应789到385Hz的电磁波频率）范围内能被人看到。在物理学上，"光"的概念则指所有波长的电磁波
2. 美国航天局太空照片，30s慢拍摄影，多张照片的重叠影像，可以看到城市光影的轨迹
3. 太阳表面
4. 埃及浮雕阿玛纳时代（公元前14—13世纪）：奈费尔提蒂祭拜太阳。图片：DETAIL Zeitschrift fuer Architektur Serie, 2002.6
5. 巴塞罗那圣家族教堂　建筑师：安东尼·高迪

1. Usually the light can be seen in the range of the electromagnetic wavelength from 380 to 780 nm (corresponding electromagnetic wave frequency of 789 to 385 Hz). In physics, the term of "light" refers to electromagnetic radiation of any wavelength
2. NASA Space photos. Trace of light is visible by multiple photos overlaid at 30-seconds exposure
3. Surface of the sun
4. Egyptian bas-relief of Amarna Period (BC 14-13 century): Nefertiti worshipping the sun. Picture: DETAIL Zeitschrift fuer Architektur Serie, 2002.6
5. Sagrada Familian, Barcelona. Architect: Antonio Gaudi

"光源"一词于希腊语"Phos"，是指明亮的意思。光是地球生命存在不可缺少的元素。光是人类生活的依据。光是温暖、是希望。使用和挖掘自然光的价值极限，是建筑师应当追求的。如何让自然光在一个进深超常的办公室照射到位？只要通过一个有一定技术含量的反光板即可把光线带入室内日光不可及的地方。

优化自然光

大楼的自然照明，特别是人们每天工作的办公空间，需要更多的自然光线，遮阳控制、防眩光控制是构筑高品质办公空间的重要因素。现代技术使自然光的使用完全可以通过大楼的控制系统操控。在莱茵集团总部大楼项目中，同层高相同尺寸的外墙玻璃面的设计就最大限度地使用了自然光。其中鱼嘴外墙构件使自然光尽可能地进入室内，同时金属的集合吊顶又反射了自然光，使日光照射不到的区域尽可能多地得到自然光线。不同的日光强度通过配套设计的遮阳帘和半透明的防眩光帘控制和调节，满足了人们对舒适的要求。

太阳能源

太阳主要是由氢（71%）、氦（27%）、重元素（2%）构成的，太阳从中心向外可分核反应区、轻射区、对流区和太阳大气层。太阳的大气层像地球一样，按不同高度和不同性质分成各个圈层，即光球、色球和日冕三层。我们平常看到的是太阳的表面，是太阳大气的最高层，温度大约为6000℃。太阳核心区的温度极高，约$1.5×10^7℃$。压力大约在$2×10^{11}$bar，使得氢的热核反应得以发生。从而释放出极大的能量。这些能量再通过辐射层和对流层中的物质传递，才传递到太阳光球底部，并通过光球向外辐射出去。氢的同期每1克的能量大约在$1.7×10^5$kW·h，剩余的太阳照射总数约为1300W/㎡，在世界各地可达到1000 W/㎡。

太阳能照射

太阳光线的直接照射是在没有散射的情况下，直接到达地球表面，它的大部分能量可以被太阳能设备接收。非直接的太阳照射将在进入大气层时被灰尘和其他物质阻碍，造成不同角度的散射，继而到达地球表面。

人工照明

人工照明的设计应该顾及使用者的心理感受，同时要根据不同的工作性质来进行设计，目的是希望得到近似日光的舒适光线。在汉莎航空公司总部大楼的吊

LIGHT

2

顶灯光设计中,建筑师汇集了白天和夜晚所需要的全部灯光强度数据,并考虑到计算机显示屏的使用,采取了防止眩光的措施,并使用了弱光灯,设计的三个强灯光则用于夜间或黄昏时期的照明。这些灯光系统由专业的灯光生产企业使用现代工艺生产。

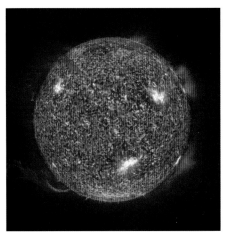

3

6. 日光的照射系数,日光的照明强度在空间通过系数反映:
   1-日光到达室内空间中部的系数 2-日光到达内墙的系数 3-墙体和屋顶均被照亮
7. 德国斯图加特火车站项目—光眼 建筑师:英恩霍文建筑师事务所

6. Daylight Coefficient, the daylight illuminance in space is measured by daylight coefficient:
   1-Daylight coefficients at the middle of interior space 2-Daylight coefficients by the interior wall
   3-The walls and the ceiling are lighted
7. 21st Century Railway Station- Light eye. Architects: Ingenhoven Architects

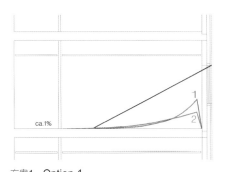 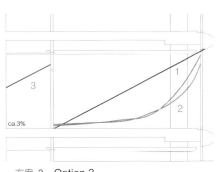

方案1 Option 1   方案2 Option 2
传统窗 Traditional Glass Window   现代全玻璃幕墙 Morden Glass Façade

4

7

5

LIGHT

8
8. 汉莎航空公司总部大楼办公室——自然光应用
9. 深圳招商酒店　建筑师：中国建筑技术集团有限公司
10. 麦加Haram清真寺城市广场遮阳广场设计　建筑师：英恩霍文建筑师事务所
11. 博鳌大庆酒店遮阳廊　建筑师：中国建筑技术集团有限公司
12. 贵阳人才大厦玻璃幕墙建筑　建筑师：德雅视界建筑事务所
13. 集合吊顶局部图

8. Lufthansa AG Headquarter, Frankfurt - Natural light office
9. Shenzhen Merchants Hotel - Glass Façade. Architects: CBTGC
10. Al Haram Mosque, Mecca. Architects: Ingenhoven Architects
11. Boao Daqing Hotel - Sun protection gallery. Architects: CBTGC
12. Guiyang Job Center. Architects: Ideal Architects
13. Detailed drawing of dropped ceiling

Light is the essential energy source for life on earth. Light is warm, light is hope. Architects must exploit and excavate natural light to the extreme.
How to let natural light reach an office significant in depth? It can be done by simply using a reflective plate to direct the daylight to illuminate indoor as the window light does.

Natural Light Optimization

More natural light is favorable for indoor lighting of a building, especially in daily working area. Sunscreen and anti-glare control is an important factor to build a high-quality office space. Modern technology makes it possible to regulate natural lighting within the building. The design of floor-to-ceiling glazing of the RWE headquarters maximizes the usage of natural light. The fish-mouth of the façade allows natural light to reach indoor as much as possible, and the metal ceiling assembly reflects to light the area without direct daylight. Daylight intensity can be regulated to any comfortable level by adjusting the sunscreen and the translucent anti-glare curtain.

Solar Energy

The sun is composed primarily of hydrogen (71%), helium (27%), and the rest are heavier elements (2%). From the center outward are the nuclear reactive zone, radioactive zone, the convective zone, and the atmosphere. The Earth's atmosphere is divided into layers of different height and nature. Similarly, the solar atmosphere consists of the photosphere, the chromosphere and the corona. The solar surface visible to the naked eye is the bottom of the atmosphere, with a temperature of 6,000 deg C. The solar core has a temperature of 15 million deg C and a pressure of 200 billion Bar, enabling nuclear fusion reaction. This process coverts hydrogen into helium, producing tremendous thermal energy. The energy must travel through the radioactive zone and the convective zone to the bottom of photosphere, before it radiates into the space.

Solar Irradiation

When the direct irradiation of the sun rays reaches the Earth's surface without scattering, the majority of its energy may be received by solar devices. Non-direct irradiation refers the sunlight hampered by dust and other substances when it reached into the atmosphere, causing scattering in different angles and then to the Earth's surface.

Artificial Lighting

Artificial lighting design should take into account the user's psychological feeling depending on various working conditions. The desired effect is as comfortable as daylight.

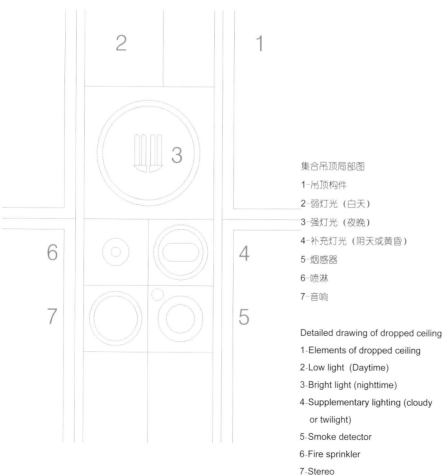

集合吊顶局部图
1-吊顶构件
2-弱灯光（白天）
3-强灯光（夜晚）
4-补充灯光（阴天或黄昏）
5-烟感器
6-喷淋
7-音响

Detailed drawing of dropped ceiling
1-Elements of dropped ceiling
2-Low light (Daytime)
3-Bright light (nighttime)
4-Supplementary lighting (cloudy or twilight)
5-Smoke detector
6-Fire sprinkler
7-Stereo

项目解读：上海科技馆

建 筑 师：英恩霍文建筑师事务所

CASE STUDY: SHANGHAI SCIENCE TECHNOLOGY MUSEUM

ARCHITECTS: INGENHOVEN ARCHITECTS

SHANGHAI SCIENCE TECHNOLOGY MUSEUM

上海科技馆项目位于上海浦东，占地面积68 000 ㎡，总建筑面积为98 000 ㎡，1996年举行国际招标，德国建筑师团队克里斯多夫·英恩霍文被邀请参加并最终获得三等奖。我作为建筑师团队的一员，这是继上海世贸国际广场项目之后参加的又一个中国项目。克里斯多夫·英恩霍文团队设计构思简单、明确，用现代的建筑技术来体现科技馆的内涵。展览厅呈月牙形，弧形的东北面用自然石材表达了中国长城的含义和中国古代的科技发展，而在东南面的大面积玻璃幕墙及玻璃屋顶则展现了中国今日的科技水平。全玻璃屋顶使每个展厅直接或间接地接受到自然的光线，屋顶设置了遮阳板来阻挡夏日阳光的照射，遮阳板同时是太阳能接收板，其生产的电力可供给科技馆的部分需求。

科技馆由地下一层和地上四层构成，并附带一个相同造型的办公楼，科技馆设有地质、生物、信息、医学、航天等十多个展区，还附带了特种影院、普通影院、四维影院及太空影院，多方位、多视角地展示了现代科技的发展水平。

Shanghai Science & Technology Museum covers an area of 68,000 square meters in Pudong, Shanghai, with a floor space of 98,000 square meters. Christopher Ingenhoven Architects was invited to the 1996 international tender, and won the third prize in the final. It was the second Chinese project I participated as an architect of the design team after the project of Shanghai Shimao International Plaza. Simply and explicitly, the design used modern construction technology to manifest the connotation of the science & technology museum. The northeast side of the crescent-shaped museum is finished with natural stone, symbolizing the Great Wall and the achievement of science and technology of ancient China. The glass facade of the southwest side and the glass roof represent the progress of science and technology nowadays. The glass roofing allows direct or indirect natural lighting in every exhibition hall. The sunscreen on the roof is made of solar panels, generating electricity to supplement the power supply for the museum.

The science & technology museum has one level underground and four floors above ground, accompanied by an office building modeled similarly. The museum has opened more than ten exhibition halls covering geology, biology, information technology, medical science, aerospace science, and so on. It is also equipped with a special theater, a movie theater, a 4-D movie theater, and a space theater. The museum is dedicated to show the development of modern science and technology from multiple aspects and perspectives.

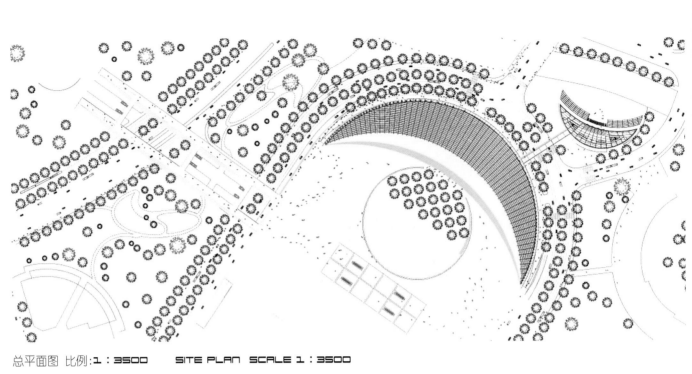

总平面图 比例：1：3500　　SITE PLAN SCALE 1：3500

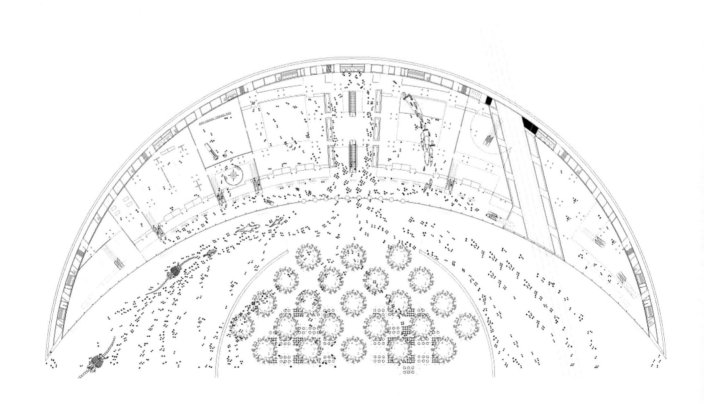

一层平面图 比例：1：1500　　GROUND FLOOR SCALE 1：1500

ㄱ 空气
AIR

空气是人类生活中最重要的元素之一。虽然科学家们对空气的研究成绩显著，但建筑师对通风的研究却经历了漫长的历程。借助于现代科技的成果，建筑师们关于空气对建筑影响的研究也逐渐成熟。

AIR PLAYS ANOTHER CRUCIAL ROLE IN HUMAN LIFE. THOUGH SCIENTISTS HAVE ACHIEVED A GREAT DEAL IN STUDYING IT, ARCHITECTS WERE SO LUCKY DESPITE THEIR LABORIOUS EFFORTS IN VENTILATION RESEARCH. THANKS FOR THE LATEST DEVELOPMENT IN SCIENCE AND TECHNOLOGY, ARCHITECTS' KNOWLEDGE ABOUT HOW AIR CONTRIBUTES TO BUILDINGS IS GROWING AT AN UNPRECEDENTED SPEED.

空气
AIR

1. 空气
2. 杜塞尔多夫城市银行外墙通风窗　建筑师：英恩霍文建筑师事务所
3. 北京普天首信新办公楼通风中庭
4. 莱茵集团总部大楼外墙通风状态 ——进风
5. 莱茵集团总部大楼外墙通风状态 ——排风
6. 影响风速的基本元素是来自这一地区风速的风数式及地理因素。图例展示了莱茵集团总部大楼风速状态
7. 普通窗全开状态，空气交换量每小时数据
8. 全玻璃窗全开状态，空气交换量每小时数据

1. Air
2. Stadtsparkasse Dusseldorf Headquarters glass window. Architects: Ingenhoven Architects
3. Beijing Putian Headquarter
4. RWE Headquarter Building's exterior ventilation - air supply
5. RWE Headquarter Building's exterior ventilation - air exhaust
6. The basic elements affecting the wind speed are the fractional extraction of the wind and the regional geographical factors. The schematic diagram shows the wind condition of RWE building
7. Window RWE Group Headquarter Building's exterior ventilation - wind exhaustion
8. Glass RWE Group Headquarter Building's exterior ventilation - wind intake

　　"空气"一词源于希腊语"aer"。空气是指人和动物呼吸的物质。空气是我们人类生活中最重要的元素之一。虽然科学家们对空气的研究成绩显著，但建筑师对通风的研究却经历了漫长的历程。依据现代科技成果，建筑师们对空气对大楼的影响的研究也逐渐成熟。建筑师们尝试在没有辅助设备的情况下，在高层建筑中打开窗户进行自然通风，并通过研究，发展高技术性的外墙来优化建筑的通风系统。

空气动力的优化

　　气流客观上是不可以调节的，但可以通过技术手段加以控制。在建筑中通过对气流的控制，调节室内温度，让室内达到人们主观需要的舒适度。

　　莱茵集团总部大楼是新生代第一个真正意义上的节能大楼。大楼的外墙体系是双层玻璃幕墙构造，其作为构件第一次应用于高层建筑中。双层玻璃幕墙是由外层玻璃、内层玻璃、通风构件和空气廊构成的。位于外层玻璃和内层玻璃间的空气廊起到了隔热保温和阻挡风压的作用。

风洞研究

　　为了减小风压给莱茵集团总部大楼带来的影响，建筑师和工程师在分析了噪声、辐射等不良因素对高层建筑的损害后，以1：500的模型进行了各项实验。

　　在一个1：64的模型中进行了空气对流的实验，实验的位置位于外墙和内墙间的空气廊，在实验中分析了空气的对流状态及强风压和排风在一个标准层的状态。

　　在1：1的外墙模型中分析了风声、风压及外墙通风口，并对其结果进行了优化。

层间的对流

　　通过计算机模拟试验将层间自然的气流在风洞里进行实验，比如在风速4m/s状态下对各种开窗状态进行观察和试验，并分析了办公室门、走廊门在不同程度打开的状态。为了限制风压对大楼的影响，建筑师进行了两个方案的分析对比，一个是双层幕墙，另一个则是普通的单层玻璃外墙，其结果完全不同。每一幢建筑、每一项任务、每一处节点都将带来新的认识和预想不到的可能。

2

3

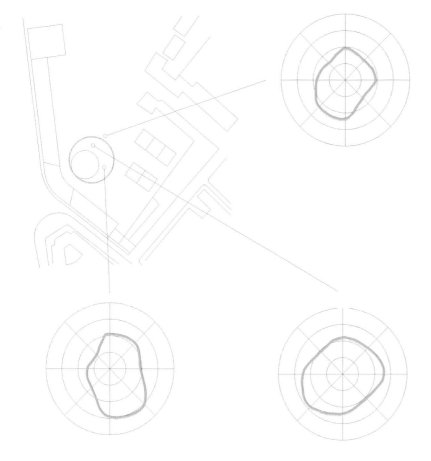

4

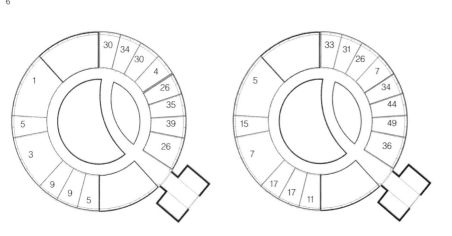

6

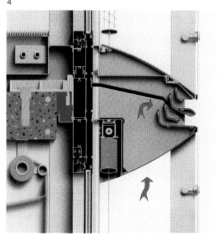

5

7 8

AIR

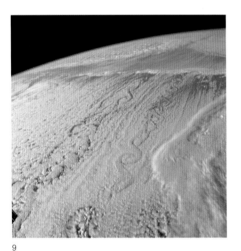

9
9. 太空照片

9. Space photos

Air, derived from the Greek "aer", refers to the material for people and animals to breathe with. Air is the most important element in our human life. Scientists have achieved a great deal on the research of air, while architects have gone a long way studying ventilation. With the development of modern science and technology, the examination on the impact of air to buildings has gradually matured.

Aerodynamic Optimization
Objectively speaking we do not have control over the atmosphere, but we can technically regulate the air flow in the building, adjusting the indoor temperature.
The RWE headquarter building is the first energy-saving building of the new era in true sense. Double glass façade had never been applied to high rise before. The double glazing consists of both the outer and inner layer glass, the air gallery between and ventilation components, which blocks the wind pressure and servers indoor climate regulation.

Wind Tunnel Research
On the scale of 1：500, the architects and the engineers conducted various experiments to study the impact of wind pressure on the RWE headquarter building, and to analysis any possible damage to the high rise from noise and radiation.
On the scale of 1：64, the experiments targeted the air convection in the air gallery, as well as the strong wind press and ventilation in one standard level.
On the scale of 1：1, a facade model is used to analyze wind noise, wind press as well as the ventilation component. The design was optimized accordingly.

Convection between Floors
The experiments of air flow convection between floors in the wind tunnel were simulated on computer. At a wind speed of 4m/s, with the windows as well as the door of the offices and corridors open to different degree, the air flow patterns were observed and recorded. To minimize the impact of wind pressure on the building, the architects compared two design variants - double layer facade vs. ordinary single layer facade, and the results are completely different. Every building, every task, and every detail may lead to new knowledge and unexpected possibility.

悉尼布莱街1号大厦　建筑师：英恩霍文建筑师事务所

这一139m高的建筑于2011年建造完成，紧凑的几何形状和旋转的建筑形体成就了建筑所需的功能要求和利益，所有的办公室都将看到悉尼海港的美丽景观。

门厅处设计了公共广场，这一高楼成为城市天际线的标志建筑。大厦不仅需要有较高的空间利用率，更要体现生态设计，这是独一无二的建筑，是为悉尼和澳洲大陆特别创造的。

该建筑获得Six- Star World Leadership的认证及Australian Green Star节能大奖。这一建筑为双层玻璃幕墙系统，大楼通过中庭实现自然通风。

1 Bligh, Sydney　　Architects: Ingenhoven Architects

The 139-metre tall 1 Bligh office building in Sydney, Australia, was inaugurated in autumn 2011. Due to its compact geometry and a slight rotation of the building in relation to its site, all offices enjoy unobstructed views of the city s beautiful harbour. All the typical functions of the ground floor have been raised to allow a public plaza at street level. The high-rise building is a new highlight in Sydney s skyline.

In addition to its spatial efficiency the building boasts an ecological concept that is not only unique for Sydney but also for the entire Australian continent.

The building is the first high-rise tower to rceive the highest six star certificate of the Australian Green Star enviromental rating system. Its fullyglazed tower is equipped with a double-skin façade and ventilated by an atrium that rises the building s entire height.

项目解读：合肥招商银行大厦
建 筑 师：郭小平
CASE STUDY: MERCHANTS BANK BUILDING, HEFEI
ARCHITECT: XIAOPING GUO

MERCHANTS BANK BUILDING, HEFEI

塑造一个城市的标志、提升城市形象成为招商银行大厦的设计主题。基地位于合肥市中心地带，临近杏花公园。主楼造型设计是这一建筑的灵魂，基本形是一个扭转的几何体，4个基本形组合构成了单体建筑。

这一单体造型体现了建筑的力量和连接，它将成为城市标志，并冲击着每个人的视觉。建筑分三部分组成：主楼39层，高180m，为办公楼；辅楼20层，高98m，为公寓楼；裙房3层，高18m，为银行营业厅及辅助商业楼。

To create a landmark and to promote the city image, it is the keynote for the design of the Merchants Bank Building. The building lot is located in the heart of Hefei City, near Xinghua Park. The design of the main building is the soul of our construction. The basic model is a geometry changing in direction, defining the prototype of the body. Four prototypes combine into a uni-body, embodying strength and connection.

Radiating vision impact, this uni-body building would be the city's new landmark. The building is divided into three parts: the 39-storey 180 meters high-rise main building for office, the 20-storey 98 meters auxiliary building for apartment, and the 3-storey 18 meters adjoining building for banking and commerce.

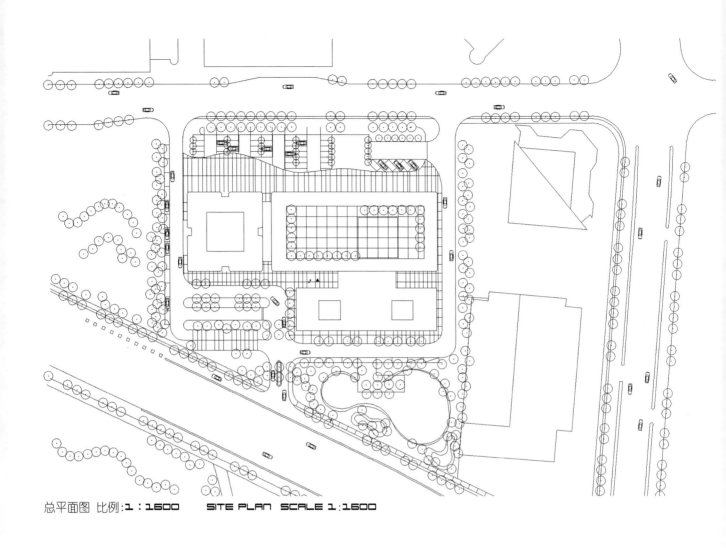

总平面图 比例：1：1600   SITE PLAN SCALE 1:1600

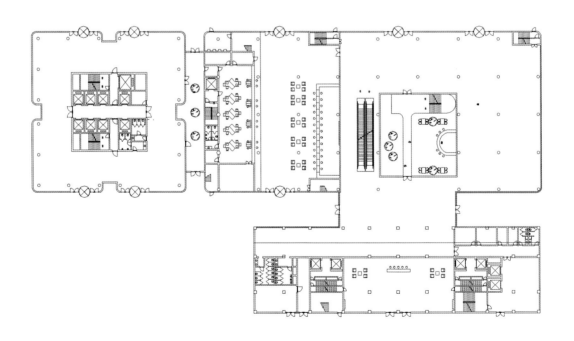

一层平面图 比例：1：800   GROUND FLOOR SCALE 1:800

项目解读：德国慕尼黑UPTOWN建筑群
建 筑 师：英恩霍文建筑师事务所
CASE STUDY: UPTOWN MUNICH, GERMANY
ARCHITECTS: INGENHOVEN ARCHITECTS

UPTOWN MUNICH, GERMANY

UPTOWN建筑群是由一座高146 m的方形塔楼和6个U形7层办公楼构成的，总面积为84 000㎡，位于慕尼黑奥林匹克体育场附近。这一项目在规划过程中，建筑师提供了三个不同高度的方案：220m、180m、146m。美国"9·11"事件的发生使高层建筑在德国被重新讨论和认识。经过多方论证，慕尼黑市政厅最终批准了高146m的方案。（未来慕尼黑的城市建筑高度将不允许超过UPTOWN大楼）

UPTOWN办公建筑群经历了十年的规划和论证，建筑师协同工程师、能源专家、幕墙专家、材料专家等，对方案进行了多次优化，把革命性的技术运用到这一项目中，使其成为未来单层幕墙节能大楼的典范。这一项目的发展过程，是从双层呼吸玻璃幕墙到单层呼吸玻璃幕墙的优化过程，自然的通风系统是由通风构件完成的，并由计算机控制开启和关闭。新型遮阳保护玻璃代替了遮阳帘，并高效地将阳光阻挡在窗外，同时起到保温的作用，使热量及冷量不易流失，降低了整体造价，单层呼吸玻璃幕墙成为未来玻璃幕墙的发展方向，它将引领着未来高科技幕墙技术的发展。

建筑师协同工程师优化了"成熟"的设计方案，为符合和创造新的国际标准，不惜代价地寻找最好的方法和最优的答案。

The Uptown building group is constituted of a high-rise square towers (146 m) and 6 U-shaped 7-story office buildings, with a total area of 84,000 square meters, located near the Olympic Stadium in Munich. The architects originally proposed solutions of three different heights, 220m, 180m, or 146 m. Yet the "9·11" terrorist attack in the U.S. brought renewed discussion and awareness of the risk of high-rise buildings. After many discussions and investigations, the Munich City Hall finally approved 146m for the building height. (In the future no construction is allowed to exceed the Uptown Tower in height within Munich city limit.)

The UPTOWN office buildings experienced a decade of planning and demonstration. Architects collaborated with engineers, energy experts, curtain wall experts and material experts to optimize the design many times and applied revolutionary technologies to this project. It became a model for future energy-saving buildings using single-layer curtain wall façade. The project took on the optimization from double-layer to monolayer breathing glass curtain wall. The natural ventilation system is completed by the venting component controlled by the computer. Instead of sunscreen curtains, the new sunscreen glass was used to block the sunlight efficiently and functions as thermal insulation, reducing the overall building cost. Monolayer breathing glass pointed out the direction of glass curtain wall, leading the development of future high-tech curtain wall technology.

The architects proposed the building design and collaborated with the engineers to optimize a "mature" product. To comply with international standards or to create a new one, they sought the best path to the best solution at all cost.

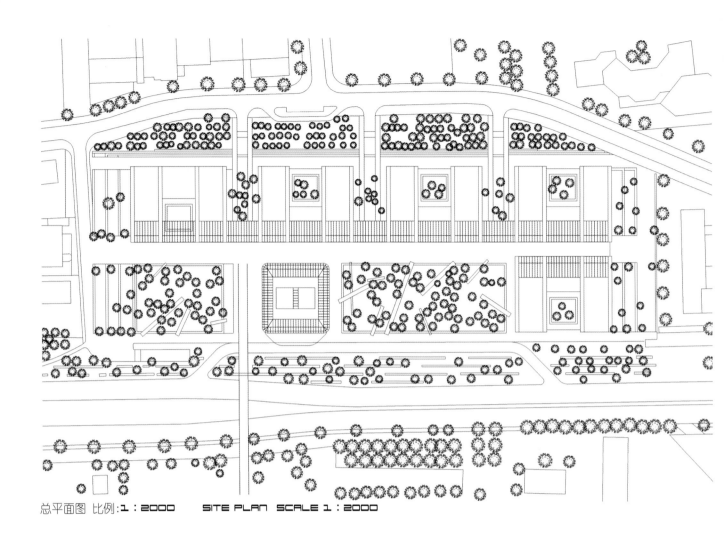

总平面图 比例：1：2000　　SITE PLAN　SCALE 1：2000

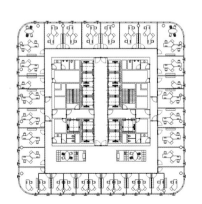 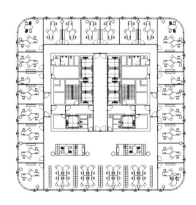 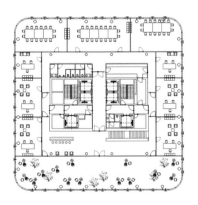

平面图 比例：1：800　　PLANS　SCALE 1：800

8 集合作用
SYNERGY

建筑师设计了建筑方案,并协同工程师优化了"成熟"的产品。我们努力地满足多方面的要求和国际标准,寻求解决问题的最好办法,寻求独立的答案,在其中我们耗费了空间、时间、资金和资源。

ARCHITECTS PRODUCE DESIGNS AND COLLABORATE WITH ENGINEERS IN PERFECTING THESE "DEVELOPED" PRODUCTS. THEY SEEK TO SATISFY MULTIPLE DEMANDS AND REQUIREMENTS, AS WELL AS INTERNATIONAL STANDARDS, PERFECT SOLUTIONS TO PROBLEMS, AND SEARCH INDEPENDENTLY FOR ANSWER, DURING WHICH SPACE, TIME, MONEY AND RESOURCE ARE CONSUMED.

8 集合作用
SYNERGY

1. 欧洲投资银行玻璃屋顶的设计不仅仅包含了轻巧的结构外壳的塑造，更多的是集合了众多的功能需求，如中庭通风、灯光照明、消防装置等。
2. 集合吊顶图解
3. 汉莎航空公司总部大楼集合吊顶
4. 悉尼布莱街1号大厦中庭

1. The design of the glass roof of the European Investment Bank includes not only the shape of the lightweight structure of the shell but more the collection of functional requirements, such as atrium ventilation, lighting and fire control installations
2. Congregation Ceiling Diagram
3. Detail of ceiling assembly piece used in RWE Headquarter Building
4. The atrium of 1 Bligt Street, Sydney

"集合作用"一词来源于古希腊语"Synergia"。集合作用是"协同工作"的意思。

一种形式的综合作用是由物质材料和各种因素决定的。

建筑构件的集合作用

现代建筑是由众多建筑构件组合而成的，每个构件又是由各种功能组件组合而成的。这种由各种功能组件组合的构件我们称之为"集合"构件。集合构件的出现使各种构件的数量明显减少，这不仅仅出于经济的原因（减少原材料的使用，降低能耗，缩短生产时间，降低成本），也是优化了工作程序和承担了保护生态责任的结果。莱茵集团总部大楼的多功能"集合"吊顶是在基于环保、高效、精工等因素下设计而成的。它是一个被集合的建筑构件，它统一了所有的技术设备，如：电源接口（灯光）、数据线、烟感器、对讲机、扬声器、冷却器（冷却吊顶）、喷淋等。它们被统一集合在一个模板上，通过工业产品设计师的设计组合，最终完成的是一个做工精良的集合产品。

例如为了优化冷却吊顶部分的功效，建筑师通过一个扩大的吊顶表面，将冷水管安装在冷却百叶中，冷却口使用了隔热材料，同时起到了吸声的作用。在吊顶集合构件里还组合了可以转换的灯光系统，使人们在没有眩光的环境里，面对计算机显示屏工作，同时又补充了自然光在阴天照明不足的问题。吊顶集合构件是根据静力学的理论，使吊顶尺寸尽可能减少并且自由悬挂在钢筋混凝土楼板上，目的是利用钢筋混凝土楼板与集合吊顶的间隙，使空气循环，达到热量排出的目的。

时刻准备着改变你的观念和位置，时刻准备着各种可能、各种解决办法。

Synergy comes from the Greek word "synergia", meaning "joint work and cooperative action".

Synergy of any form is determined by the material substances and other factors.

Synergy of Building Components

The modern architecture is composed of many building parts. Each part in turn is assembled by various functional groups of components, which we call "synergic" components. The emergence of synergic components significantly reduced the number of various assemblies. Not only is it driven by economic reasons (to reduce raw material usage, energy consumption n, production time and the overall costs), but

SYNERGY

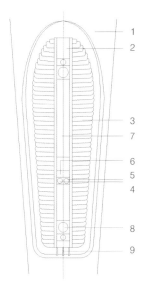

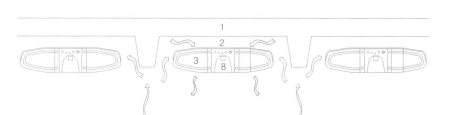

集合吊顶图解	Congregation Ceiling Diagram
1-钢筋混凝土楼板	1-Reinforced concrete plate
2-技术线路	2-Cechnical lines
3-冷却叶和吸声板	3-Cooling blades and sound-absorbing board
4-音响	4-Stereo system
5-烟感器	5-Smoke detector
6-喷淋	6-Sprinkler
7-弱灯光（白天）	7-Low light (daytime)
8-强灯光（夜晚）	8-Bright light (nighttime)
9-连接口	9-Connecting port

2

also a result of optimizing working procedure and committing ecological environment protection. The design by industrial product engineers, this synergy component integrates all technical equipment, such as power multifunctional "synergic" ceiling of the RWE headquarter building is designed based on environmental protection, high-efficiency and sophisticated technology. A state of the art design by industrial product engineers, this synergy component integrates all technical equipment, such as power interface (lighting), data cable, smoke detector, intercommunication device, speaker, and cooling water (cooling ceiling).

In order to optimize the cooling effect, the size of the ceiling is expanded to install the cold water pipe behind the cooling louver. Insulation materials are applied to the cooling opening, functioning as acoustic absorber as well. The convertible lighting system of the ceiling component provides a computer user friendly environment free of glare, and supplements natural lighting on a cloudy day. A design based on theories of statics, the ceiling component hangs freely from the concrete floor above, leaving the gap for enhanced air cooling.

Be ready to change your idea and position; be ready for any possibility and solution.

3

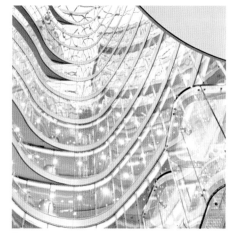

4

项目解读：湖北省输变电工程公司办公楼
建 筑 师：德雅视界建筑师事务所
CASE STUDY: OFFICE BUILDING OF HUBEI POWER TRANSMISSION COMPANY, WUHAN
ARCHITECTS: IDEAL ARCHITECTS

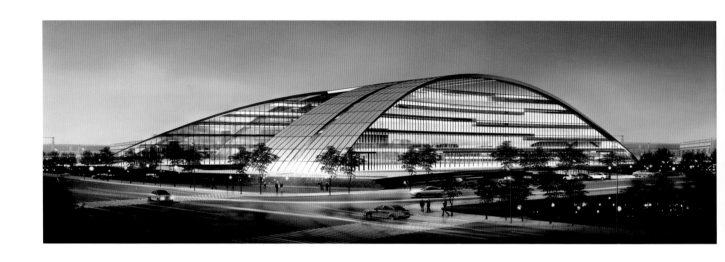

OFFICE BUILDING OF HUBEI POWER TRANSMISSION COMPANY, WUHAN

湖北省输变电工程公司办公楼设计的主题是创造一个现代、时尚的建筑，以展现当代中国国企开放和积极的企业文化，并在武汉新区塑造新的城市形象。

该方案中，地理条件决定了设计的轨迹：景观和建筑设计遵循了原有的地理条件，利用周围的水系创造具有风景区的办公空间，形成了新的文化景观，并同周围自然水系完美融合。

该建筑单体平面呈H形，两翼为办公楼，中心部分为中庭，两翼为弧形的造型，屋顶直接与地面连接，成为建筑的第五立面。利用原有的水系塑造了同建筑弧形风格相同造型的水系，使北侧建筑置身于水景当中，建筑在水中形成的倒影，增添了美感。两翼间的半开放空间为现代中式庭院。北侧为主办公楼，共六层，每层中轴设有四个弧形楼梯连接各层，从而提高部门间的工作效率。

中庭位于两个弧形建筑之间，高六层。它是一个公共的活动空间并赋有多种功能，可举办大会、公司庆典等。中庭四层高处设有弧形天桥以连接两个建筑，从而提高工作效率。中庭具有自然通风系统和高效遮阳构件，成就了零能源中庭建筑。南侧弧形建筑是根据甲方要求设计的有三个独立入口的办公楼，每个入口为两个单元共用，每个单元都拥有独立的电梯和弧形楼梯。景观设计为园林式办公提供了基础，并成就了风景区办公的模式。

The design theme of office building for Hubei Power Transmission Company is to create a modern, stylish building to represent the open and positive corporate culture of modern Chinese state-owned enterprises, and to establish a new image of the city in the new district of Wuhan.

The design integrates various building systems to create a new urban landscape. Geographical conditions constitute the trajectory of our design: the landscape and architectural design of comforts to the geographical conditions, creating a scenic office area blended perfectly with the surrounding water system. The building complex has the H-shaped layout: two wings as the office, the center as the atrium. The two wings are of the arc-shape, and the roof extends to the ground, becoming the fifth facade of the building. The original water system is transformed to match the curved style of the architecture. The image of the north side building immerges into the waterscape, and its reflection adds beautiful touch to the scene. The semi-open space between the two wings forms a modern Chinese courtyard. The architectural layout is defined by both the arc-buildings at the north and south side, connected by the atrium. The north side is the main office building with six stories. At the central axis, there are four curved staircases connecting each floor to improve inter-department collaboration.

The atrium is located between the two arc-shaped buildings. It is a public and multi-functional space for activities such as general assembly and company celebration. A curved overpass bridge connects the two buildings to improve the working efficiency. The atrium accomplishes zero energy design, thanks to natural ventilation and high-efficient sun-block technology. The arc-building on the south side has three separate office entrances by client's request. Each entrance is shared by two companies, and each unit has its own elevator and curved staircase. The landscaping design provides the basis of a garden-style office building and creates a model of scenic office complex.

总平面图 比例:1:2000　　SITE PLAN SCALE 1:2000

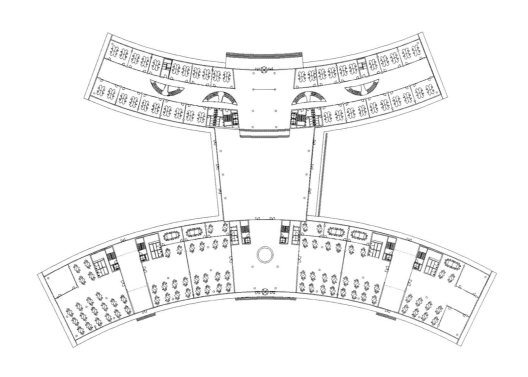

一层平面图 比例:1:1500　　GROUND FLOOR SCALE 1:1500

9 工业产品与建筑
INDUSTRIAL PRODUCTS AND ARCHITECTURE

新建筑的要求不再是以往的钢筋水泥浇筑及装饰材料的安装，而是工业产品。在工厂内被高精度加工的建筑构件，通过运输在现场安装完成，目的是要创建环保、经济、节能和高品质的现代建筑。

THE NEW CONCEPT OF BUILDING IS NO LONGER A CONVENTIONAL POURING OF REINFORCED CONCRETE OR A SIMPLE INSTALLATION OF DECORATIVE MATERIALS. IT HAS BECOME A REALITY THAT AS INDUSTRIAL PRODUCTS, CONSTRUCTION COMPONENTS ARE MANUFACTURED BY PRECISION MACHINERY, AND THEN TRANSPORTED TO THE CONSTRUCTION SITE TO FINISH ASSEMBLY. IT AIMS TO CREATE ENVIRONMENT AND ECONOMIC FRIENDLY, ENERGY-SAVING AND HIGH-QUALITY MODERN ARCHITECTURE.

9 工业产品与建筑
INDUSTRIAL PRODUCTS AND ARCHITECTURE

1
1. 莱茵集团总部大楼是一个全产品化、系统化的建筑，建筑师和产品设计师紧密合作，发展了大楼室内外建筑构件，如双层玻璃幕墙通风构件、室内控制面板、集合吊顶、电梯轿箱、灯具等
2. 莱茵集团总部大楼产品化吊顶
3. 莱茵集团总部大楼办公室室外信息屏
4. 汉莎航空公司总部灯具　　建筑师：英恩霍文建筑师事务所

1. The RWE headquarter building is a fully productized and systematic construction, in which architects and product designers collaborated closely and developed building components for both interior and exterior, e.g. Double glazing ventilation, indoor control panel, ceiling assembly piece, elevator cab, lighting fixture, etc.
2. Congregation Ceiling of RWE Headquarter building
3. The office outdoor screen of RWE Headquarter, building
4. Light fitting of Lufthansa AG Headquarter building. Architects: Ingenhoven Architects

新时代的建筑形式

新时代的建筑过程是直接生产、加工、安装的过程。当建筑师和开发商确定了建筑方案以后，建筑师组织产品设计师共同研究发展方案，同时协同其他专业设计生产匹配的产品。建筑师和产品设计师把设计数据提供给生产商，然后监控产品的生产过程和质量，直到产品生成、运输、安装完成。工厂加工的建筑构件拥有高质量，它们根据国际工业标准被加工成形，使每个构件细节都完美无缺。

建筑师与工业产品

建筑师在设计方案的开始就必须思考工业产品与建筑的关系，这不仅是建筑设计的规则，而且是建筑设计与产品设计及工业化匹配和兼容的问题。新时代的建筑建造，要求建筑师必须具备协同工程师、能源专家、产品设计师一起规划项目、协调各专业共同完成整体项目的能力，并共同承担最终的责任。

21世纪的发展商对建筑师有着更多、更高的要求，它们希望建筑师的能力具有更大的弹性潜质，具有国际视野，从而了解世界范围内产品的发展趋势，熟悉生产厂家的生产效率、工艺流程情况，了解哪些厂家的产品具备最高的品质、最低的价格和最佳的售后服务信息，最终能协同工程师、产品设计师完成一个自规划设计、生产加工到运输安装的系统工程。

现代建筑系统

现代的建筑系统是一个完整而复杂的综合建筑系统，现场的混凝土浇筑仅是整体系统的一部分，现代的建筑系统意在设计一个整体的解决方案，如承重结构框架的浇筑、屋顶的制造和安装、水电设备的安装、室内构件和外墙构件的生产与安装。一个普通建筑的解决方案是通过标准件组合安装构成的。标准件通过组合构成组合件，组合件通过产品设计师的设计使之简单化，而简化的过程是系统优化过程中最重要的一环。

产品系统

未来的建筑构筑系统，除了现场的混凝土框架，其余的单元构件将在工厂的生产车间加工完成。由计算机控制的生产线能够标准化地生产和加工每个元件或单元，使误差降到最小，未来工业化的建筑单元与构件决定了最终的建筑品质。

2

产品运输系统

现代的物流系统可以高效、安全地组织产品的装载、运输、存放及送达的任务。

安装系统

现代建筑的最后环节是将建筑单元构件在工地安装完成,不仅能缩短工业化建筑系统的建造周期,也能保证工程项目借助标准化的规划程序,使产品生产和工地构筑同时进行,这样可以高效、环保、系统地建造新时代建筑。

建筑的读识,是消费也是创造。展现现代科技发展魅力的建筑,被赋予了时代精神和意义。

A New Era of Construction

The construction process of the new era is the process of direct production, processing and installation. After architects and developers define the design plan, product designers join architects to work on the product design, and collaborate with engineers of other technical fields to design and fabricate accessories or choose compatible products available. The architects and product designers will deliver design specs to the manufacturer and conduct the quality control through the production process until the products are fabricated, transported, and completely installed. Building components through this process have a high quality in accordance with international industry standards to achieve the perfection of every detail.

INDUSTRIAL PRODUCTS AND ARCHITECTURE

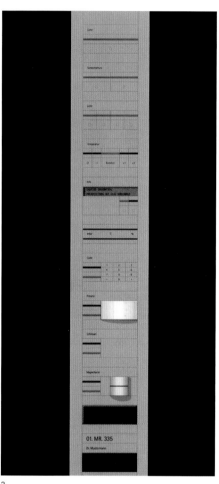

3

3. 莱茵集团总部大楼办公室室外信息屏。由上至下:灯光、遮阳保护、灯光、温度。室内状态信息如:项目会议推迟到14:00举行,压力、密码、状态、钥匙磁卡、房号、办公人员姓名

3. The office outdoor screen, from top to bottom: lighting, sun protection, lighting, temperature, Indoor status information: Meeting postponed to 14:00; pressure mbar, password, status, magnetic key card, room number, names of the office staff

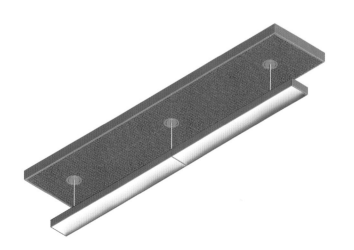

4

INDUSTRIAL PRODUCTS AND ARCHITECTURE

5.
5. 汉莎航空公司总部灯具　建筑师：英恩霍文建筑师事务所
6. 汉莎航空公司总部大堂灯具
7. 汉莎航空公司总部办公楼同样是一个全产品化、系统化的建筑，集合吊顶系统、家具系统、标志系统、由建筑师团队设计，并分别由生产商Erco Siteco Vitra生产制造
8. 莱茵集团总部大楼隔断墙
9. 汉莎航空公司总部办公楼安检装置

5. Light fitting of Lufthansa AG Headquarter building. Architects: Ingenhoven Architects
6. Lufthansa AG Headquarters lobby lighting
7. The Lufthansa headquarter building in Frankfurt is also a fully productized and systematic construction. Its ceiling assembly system, furniture system and identification system, all were designed by a team of architects and fabricated by the manufacturer ERCO Siteco Vitra
8. RWE Headquarters building partition walls
9. Security Devices for Lufthansa Headquarters Building

Architects and Industrial Products

The architects must think through the relationship between the industrial products and the building at the beginning of the design, as this is not just the rules of the architectural design, but also a systematical matter of matching and compatibility between architectural design and product design as well as it industrialization. In the new era of building construction, architects must work with engineers, energy experts and product designers together to plan the project, and coordinate the professionals of all fields to complete the whole project and jointly assume the ultimate responsibility.

The developers/clients of the 21st century expect more adaptive potential from the architects, posting more requirements and higher standards. With a global perspective, the architects should understand the developing trend of the products worldwide, be familiar with the production efficiency and process of the manufacturers to identify the best quality, the lowest price and the best customer service. Eventually the architects should be able to work with engineers and product designers to complete system engineering from design plan, processing and production, to transportation and installation.

Modern Building System

The modern building system is a complex integrated system, in which pouring concrete framework only counts for part of the work. It aims to an innovative design which provides an integrated solution, e.g., from pouring structural framework, roof manufacture and installation, water and electricity installation, to the production and assembly of interior and exterior components. One solution to common building construction is based on the composition of standard components; and in turn the synergy components are simplified by product designers from the assembly of standard building parts. So the process of simplifying is the most crucial part of our system of optimization.

Product System

In future building system all components, except for the on-site concrete framework, will be completed in manufacturer's production workshop. Controlled by the computer,

INDUSTRIAL PRODUCTS AND ARCHITECTURE

each component or part unit can be produced to the standard on the production line, so that the error can be minimized. The industrialized building components and parts will determine the final quality of building construction in the future.

Product Transportation System
The modern logistics system can efficiently and securely organize the tasks of loading, transportation, storage and delivery of the products.

Installation System
The last part of the modern building construction is to complete the installation of building components. This is not just to shorten the construction period realized by the industrialized building systems, more likely it enables the products manufacturing and construction on-site at the same time by means of standardized planning procedure, thus to efficiently, environmental consciously and systematically construct the buildings of new era.
Interpretation and understanding the architecture, is consumption as well as creation. Buildings showing the charm of the development of modern science and technology are given the significance of the spirit of the time.

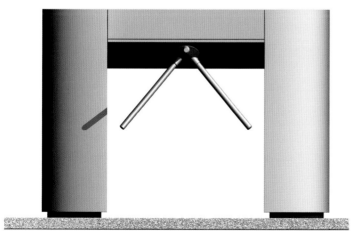

项目解读：法兰克福、巴黎、东京、底特律、日内瓦国际车展奥迪场馆
建 筑 师：英恩霍文建筑师事务所

CASE STUDY: AUDI AG PAVILION, FRANKFURT/TOKYO/DETROIT/GENEVA/PARIS

ARCHITECTS: INGENHOVEN ARCHITECTS

AUDI AG PAVILION, FRANKFURT/TOKYO/DETROIT/GENEVA/PARIS

"突破科技、启迪未来"是奥迪的广告语，在法兰克福的国际车展上，通过新颖、激情的场景展示了奥迪的技术和完美。品牌的成功和技术的突破使奥迪对场馆设计非常重视，建筑师们通过玻璃和不同色彩的灯光，展示了奥迪的内涵和技术，突出表现了奥迪百年经典的品牌魅力。

设计中根据展览馆的空间结构及白天人群运动的持续自然形态确立的设计原则：
1.创造一个逻辑的空间；
2.这个空间的自身形式能够帮助整个场馆运行；
3.建筑构件重量轻，结构简单，易于安装、拆卸和运输。

"Breakthrough technology, enlightening the future" is the slogan of Audi at the Frankfurt International Auto Show. The Audi Pavilion showed technology and perfection through a novel and passionate stage. The success of the brand marketing and the breakthroughs of technology achieved made Audi attach great importance to the venue. Through the glass and lights of different colors, the connotation and technology of Audi are clearly displayed, representing the charm of Audi's century classic brand.

The design principles are established according to the spatial structure of the pavilion and the daytime crowd distribution to:
1.Create a logical space;
2.The space structure is intrinsically conducive to run the pavilion;
3.The building components have to be light weighted, simple structured, easy to install, disassemble and transport.

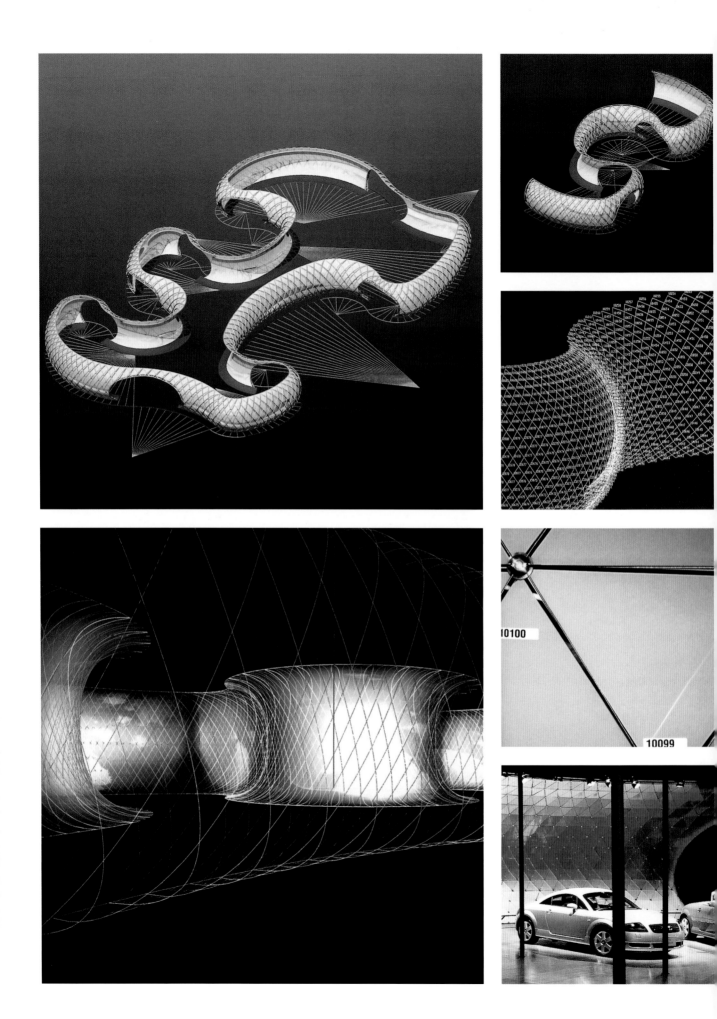

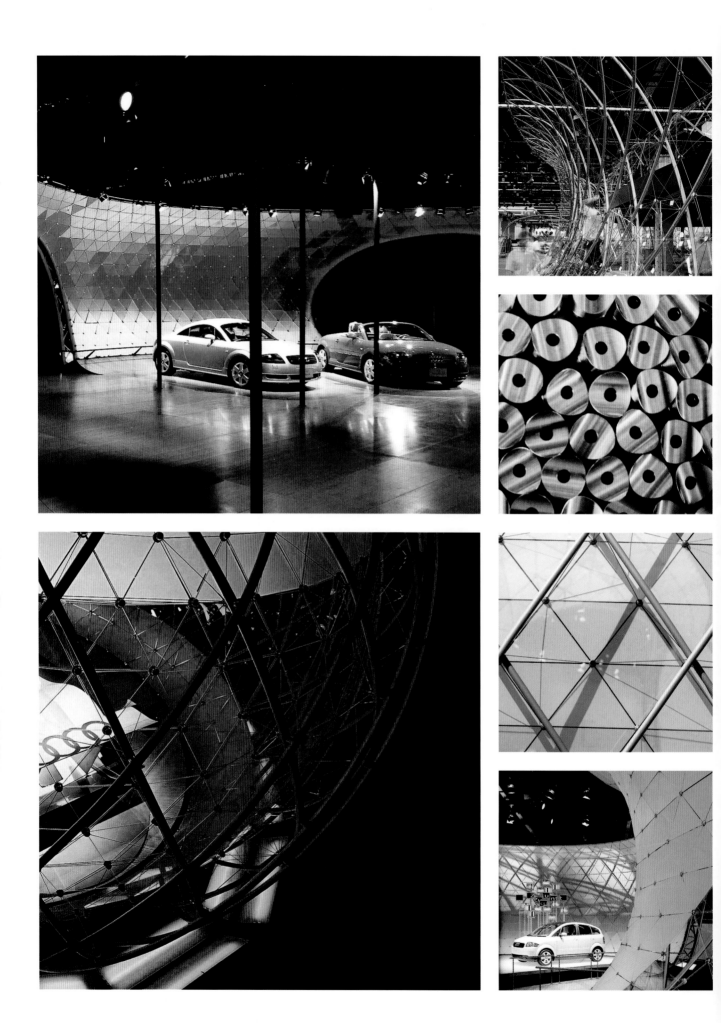

10 外墙技术
FACADE TECHNOLOGY

外墙是建筑物的名片，有逻辑性、高效的外墙设计是生态建筑的成功所在。在不受地理位置、季节和技术设备的限制下，想创造一个理想的均衡的室内温度，只有通过现代的外墙技术才可实现。巧妙的外墙设计将能源消耗降低到最小值，从而达到生态、环保和经济的目的。

The facade of the building is like the business card. A logical and efficient facade design is the key of the success of an eco-building. It relies on modern technology of exterior wall to create an ideal balance of indoor temperature regardless of geographic location, season, and technical limitations. The clever design of the exterior wall could reduce the energy consumption to a minimum, so that the ecological, environmental and economic goals can be achieved.

10 外墙技术
FACADE TECHNOLOGY

外墙的概念

从室内空间到城市空间经过的过渡层就是外墙，它有着特别的意义，对外墙的讨论成了建筑学的基础题目。外墙是建筑的一张名片（还有一种解释是观赏的外表），所以外墙在建筑设计中备受重视，一个有逻辑性高效的外墙是新时代建筑的成功所在。如果建筑师对外墙做非理性的装饰和纯粹的注重外部的视觉效果就会造成很大的危害。

外墙在古典的解释中包括了开口和墙面，现代的外墙则功能性更强，内部的活动通过外墙表现。内部和外部和谐统一，通过现代技术使大楼的外壳成为结构的帘饰。

需求

"聪明"的外墙设计可以将能源需求降低到最小值，也可以使用自然能源，降低整体成本，这样就达到了经济和环保的目的，其结果是整体受益。透明是现代建筑的一个重要特征，全玻璃建筑有一个很明显的弱点，就是冬季容易产生热量流失，而夏季则导致室内酷热。为了解决这一问题，建筑师和工程师花费了大量的时间和精力去探讨和研究，今天我们可以通过不同的技术方法解决这一问题，其中幕墙技术就是最基本的使用方法。要创造一个均衡的室内温度，而且不依赖于地区、季节和其他的辅助技术设备，只有通过包含现代技术的外墙系统才可实现。当外墙动力和转换的温度状况内外相符时，建筑的能耗是大大降低的。一个成功的外墙设计，或多或少都使用了自然能源，如太阳能、自然通风、自然采光、采热，通过技术上的解决办法将设计进行完善，既降低了综合成本，也保护了生态环境。

未来外墙系统

未来建筑将是由建筑师、结构工程师、设备工程师、建筑物理学家、能源专家、外墙专家等众多专业共同发展的产品，我们希望和需要的建筑是一个易于减轻风压的形体，使风压和外墙及大楼的承重降低到最小值。一个先进的外墙系统、模拟控制系统或数字控制系统将控制着外墙的通风、遮阳角度及大楼的技术设备，使大楼通过自然能源运行。一个精心设计的幕墙系统是新时代建筑的核心，通过高效的幕墙系统和技术设备，可以实现建筑设计的能源节约和生态保护。

The Concept of The Exterior Wall

The exterior wall is the transition from the indoor space to the urban space. Therefore, it has a special significance, and the discussion of the external wall becomes the

1

1. 外墙在古典的解释中包括了开口和墙面，阿富汗早期的砖墙建筑 图片：DETAIL Zeitschrift fuer Architektur Serie 2003, 1/2
2. 外墙记录了佛罗伦萨的历史
3. 曼哈顿的玻璃外墙
4. 莱茵集团总部大楼外墙节点轴测图
5. 上海世贸国际广场原设计外墙节点轴测图

1. Early brick building in Afghanistan. In the classical interpretation, the facade includes the opening and the wall. Picture: DETAIL Zeitschrift fuer Architektur Serie 2003, 1/2
2. Exterior wall has recorded the history of the city – Florence
3. Glass facade in Manhattan, New York
4. Isometric view of facade element RWE Headquarter Building
5. Isometric view of facade element of Shanghai Shimao International Plaza

FACADE TECHNOLOGY

2

basic topic in architecture. As the business card, or the ornamental appearance of the building, the exterior wall, also called the facade, attracts a lot of attention in architectural design. A logic and efficient facade is the key of success for buildings of new era, while any irrational decoration or solely pursuing external visual effect may largely undermine the design.

The classical interpretation of the facade included openings and walls, while the modern interpretation focused on the functional present of the internal activities and their contents.

Demand

A smart design of the facade can minimize the energy demand and reduce the overall cost by using natural energy, thus achieve both the goals of economic and environment friendly. Transparency is an important feature of modern architecture. Yet the all-glass building has an obvious weakness – prone to heat loss in winter and but heat accumulation in summer. To solve this problem, architects and engineers have spent a lot time and effort on the research to develop various technical solutions, and curtain wall is one of the basic technologies. To create a balanced indoor temperature that does not depend on the region, season and other technical equipment, the facade system with modern technology is only way to fulfill the quest. When the exterior wall establishes a dynamic balance of temperature condition between indoor and outside, the building's energy consumption is greatly reduced. A smart exterior design, using natural energy such as solar energy, natural ventilation, natural lighting and heating system, perfected through technology solutions, not only reduces the overall costs, but also protects our ecological environment.

3

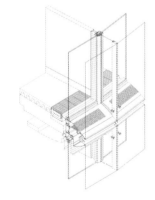

4

Future Facade System

Future buildings will be developed jointly by professionals from many fields such as architects, structural engineers, equipment engineers, building physicists, energy experts and facade experts. The shape of the building should be less subject to wind press, so that the load-bearing structure of the building and the exterior wall can be minimized. A state-of-the-art facade system, along with the analog or digital control system will regulate the ventilation, sunscreen angle, and the technical equipment of the building, so that the building can run on natural energy. The well-designed facade

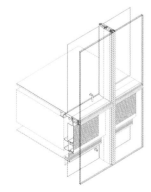

5

FACADE TECHNOLOGY

system is the core of the building in the new era, as by mean of the efficient facade technology and technical equipment, we can achieve conservation and environmental protection in the architectural design.

6. 德里斯顿办公楼
7. 合肥MV广场外墙节点
8. 现代外墙的安装 图片：DETAIL Zeitschrift fuer Architektur Serie
9. 德里斯顿办公楼外墙-由机械驱动的外置的金属遮阳帘

6. Office Building in Dresden, Germany
7. Office Building in Dresden, Germany
8. Hefei MV Plaza facades detail
9. Modern facade installation Picture: DETAIL Zeitschrift fuer Architektur Serie

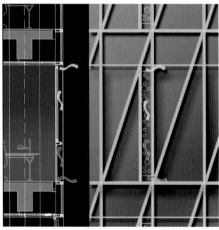

FACADE TECHNOLOGY

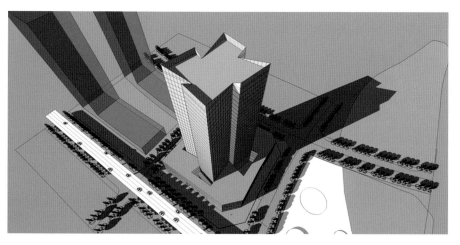

北京西部金融中心
建筑师：德雅视界建筑师事务所

Beijing West Financial Center
Architects: Ideal Architecets

项目解读：上海世茂国际广场
建 筑 师：英恩霍文建筑师事务所
CASE STUDY: SHIMAO INTERNATIONAL PLAZA, SHANGHAI
ARCHITECTS: INGENHOVEN ARCHITECTS

SHIMAO INTERNATIONAL PLAZA, SHANGHAI

上海世茂国际广场是我作为建筑师在中国参加设计的第一个项目。在这一项目中我和我们建筑师团队付出了许多心血和激情，遗憾的是由于种种原因未能按原设计实施，但它的系统是根据中国气候条件进行的，设计有其参考价值。

由于上海当地气候、空气指数、湿度、洁度不能达到呼吸幕墙的要求，所以整体系统采用封闭的双层玻璃幕墙，它同样阻挡了高空的风压和冷热气的进入。高效遮阳帘在双层玻璃之间不受风压影响，而内循环空气系统通过空气廊进行换气。上海世茂国际广场坐落于上海这座国际性商业中心的心腹地带，基地紧邻南京路，每天有两百多万顾客光顾这条中国最大、最繁华的商业大街，西面的西藏路是上海最主要的南北交通干道。所设计的外立面是双层中空玻璃幕墙，中空部分设有铝合金百叶窗，以达到高效遮阳的效果。同样，具有能量辐散作用的广告屏和可开启的内层玻璃都是根据上海地区的气候条件而设计的。

建筑外立面的特点是：建筑的承重结构通过无窗间墙的透明玻璃幕墙清楚地显现出来，钢斜撑的位置、结构、规模及其连接点，兼顾了结构承重和承载情况下的地震防御要求。玻璃幕墙本身的特色在于：它是双层玻璃幕墙，内外两层玻璃都与层高相同，因此加强了封闭与开敞空间的对比，并保证了足够的自然采光。

这层由玻璃制成的"肌肤"，以其玲珑剔透的质感，使人对其内部的骨架一览无余，建筑结构的竖直和斜向线条得以充分的显现，从而将建筑结构的表现力和幽雅的外观完美地结合起来。采用中性色调的高清晰度透明玻璃及由铝合金包裹的承重结构，使这一对比更加突出，在白天的阳光和夜晚的人工探照灯照射下，建筑物突显出它的力与美。

建筑物的能源平衡利用性和经济效益主要是以建筑物的容积率和外表面积来衡量的。在进一步的设计中，通过结构组合把建筑物的表面积控制到最小，这样可以减少夏天的热量辐射和冬天的热量消散，从而减少了能源消耗。

Shanghai Shimao International Plaza is the first project in China on which I participated as an architect. The design experience was strenuous but passionate. Unfortunately, due to various reasons the project was not implemented according to the original design, yet the systematic design adapted to the local climatic condition remained as valuable reference.

Shanghai Shimao International Plaza is located the heart land of shanghai, a metropolitan of international business center. The Nanjing Road nearby is one of China's largest and most prosperous shopping streets and patronized by more than two million customers every day. In the West is the Xizang Road, Shanghai's main north-south traffic artery. Since the local climate condition, air quality and humidity cannot meet the standard of the breathing exterior curtain wall, the whole exterior system is designed to be a close-out double glazing facade, to block the high-altitude wind pressure and the inflow of hot or cold air. This facade is made of two layers of glass curtain wall, and gap between the layers is furnished with aluminum blinds to efficiently block the sunlight. The sunscreen blinds between the double glazing are not subject to wind pressure, while the internal air circulation system ventilates through the air gallery. The facade is equipped with an energy emissive advertising screen, designed according the local climate; and the windows in the inner layer of the glass curtain wall can be opened.

The facade features the load-bearing structure of the building, exposed behind the whole piece of transparent glass curtain wall; the structure and dimension of the steel bracing and its connecting points meet both the requirements of structural load-bearing and earth quake defense. The glass curtain wall features double layer glazing; both layers are extended from floor to ceiling to enhance the contrast between closed and open space, and to ensure a high level of natural lighting.

Through the glass "skin" with the exquisitely carved texture, the framework of the building is revealed explicitly. The vertical and the diagonal lines of the building structure manifests a perfect integration of the power of the structure and the elegance of the appearance; and the contrast is highlighted by the neutral colored glass of high transparency and the load-bearing structure wrapped by aluminum alloy . The building radiates a sense of strength and beauty in sunlight, so does it under the spotlights at night.

The energy efficiency and the economic benefit of a building is related to it volume to surface area rate. Taking a step forward, we can reconfigure the space structure to minimize the surface area, thus reduce the heat irradiation in summer and the heat dissipation in winter to subside the energy consumption.

总平面模型　SITE PLAN MODEL

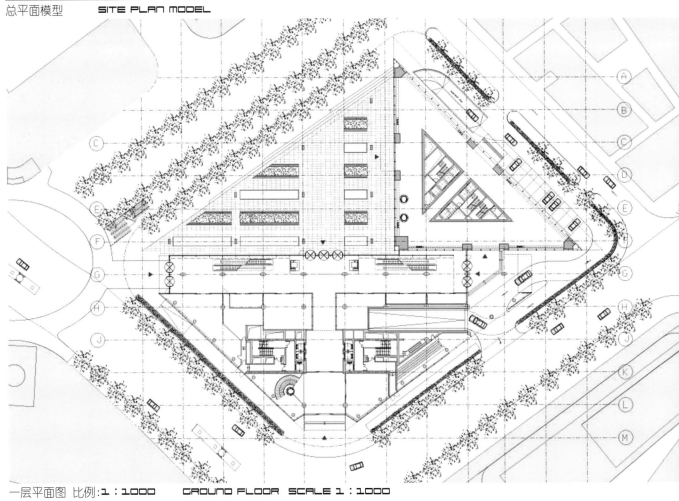

一层平面图 比例:1:1000　　GROUND FLOOR SCALE 1:1000

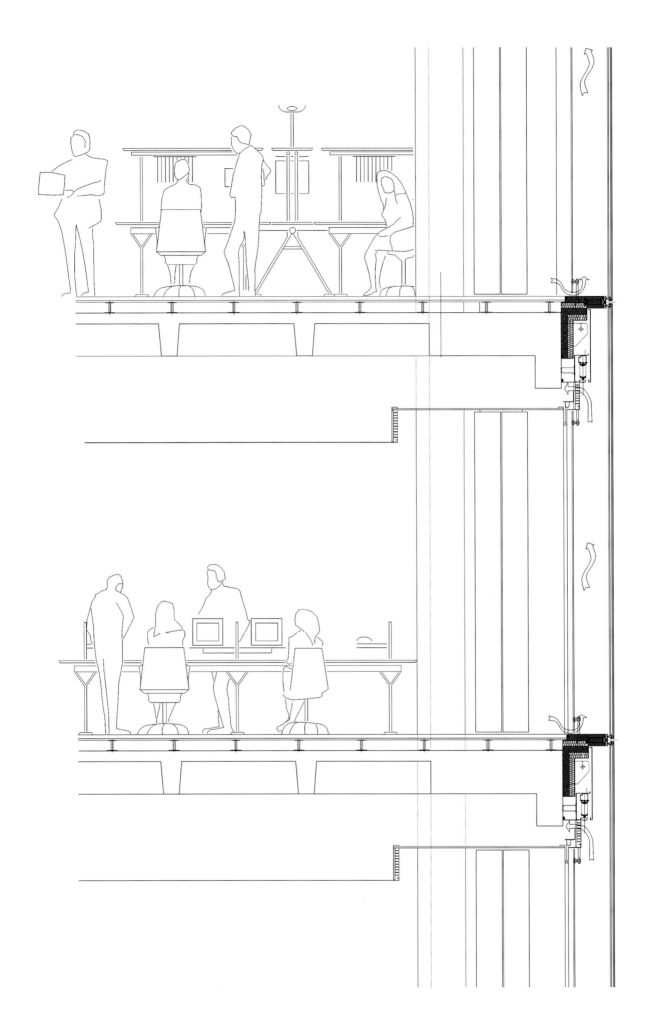

11 交流
COMMUNICATION

现代建筑应该使人们直接识别方向、目的、空间及所处的位置，并且没有其他指示及技术的帮助。空间和质量的表现必须适合空间关系。

MODERN ARCHITECTURE SHOULD LET PEOPLE BE ABLE TO DIRECTLY IDENTIFY THE DIRECTION AND FUNCTION WITHOUT OTHER INSTRUCTIONS AND TECHNICAL HELP. THE PERFORMANCE OF SPACE AND QUALITY MUST BE SUITABLE FOR SPATIAL RELATIONSHIPS.

11 交流
COMMUNICATION

现代的网络、媒介、移动资讯、导航技术虽然提高了人类信息交流的速度和效率，但也阻碍了人们之间直接的交流。

我们必须创造更多的、各种形式的空间，带给人们直接交流的可能。重视住宅的餐厅空间等于重视家庭的和谐，重视办公室内的公共空间等于重视团队的和谐，重视城市的空间等于重视城市市民的和谐。建筑师所创造的每一个空间都将承担着社会的责任和义务。

交流和引导

导向是寻找方向 —— 正确的选择方向，询问、思考，最终确定答案。引导是交流的基础，而交流是人类必要的行为，是生活的重要部分。

汉沙航空公司总部大楼的横向交通和竖向交通组织明确、直接。使使用者和来访者在大楼内可以轻松、直接地判断方向。中轴的连廊连接着办公区和中庭，在竖向的交通组织上，通过三组位于中庭的电梯将人们引入各层。电梯采用了透明玻璃电梯，无论你位于电梯的内部还是中轴廊和中庭，你都将透过全透明的玻璃，准确判断你在城市中的位置。中轴廊的月牙形空间给使用者带来了交流的可能。

1

1. 城市市场提供了人与人交流的空间和场所
2. 纽约中央公园，使人们在紧张、繁忙的工作中得到一息空气，它缓解了城市的密集和压力
3. 兰州榆中城市综合体办公楼前的城市空间给予了人们交流的机会 建筑师：德雅视界建筑师事务所
4. 杜塞尔多夫大学 建筑师：英恩霍文建筑师事务所
5. 汉莎航空公司总部办公楼中轴的交流空间 建筑师：英恩霍文建筑师事务所
6. 天津滨海科技总部区项目中心广场 建筑师：德雅视界建筑师事务所

1. The market square provides the space and place of interpersonal exchange for people
2. New York's Central Park located in the World's densest high-rise buildings, giving people a breath of air in a tense, busy working environment, and eases the dense and pressure of a city life
3. The space in front of office building gives people the opportunity to communicate - Yuzhong Urban Complex in Lanzhou. Architects: Ideal Architects
4. Oeconomicum University, Düsseldorf. Architects: Ingenhoven Architects
5. The open spaces at the central axis area of the Lufthansa AG headquarters building. Architects: Ingenhoven Architects
6. The Tianjin Binhai headquarters central plaza. Architects:Ideal Architects

7. 贵阳医学院新址 建筑师：德雅视界建筑师事务所
7. Guiyang Medical Institution Library, Guiyang. Architects:Ideal Architects

2

COMMUNICATION

3

Modern network, media, the mobile information and navigation technology improved the speed and efficiency of the information exchange, yet obstructed the direct communication between people.

We must create more various forms of space to provide the possibility of direct communication. Emphasis on residential dining space represents the valuation of the harmonic family time; paying attention to the public space of office attaches the importance to the harmonic teamwork with colleagues; appreciating the urban space is to cherish the harmony of the urban public; every and each space the architects created will have to bear the social responsibilities and obligations.

Communication and Guidance

Orientation is good for finding directions - to choose the direction correctly, asking, thinking, and making the final decision. Guidance is a basis for communication, and communication is one of necessary behaviors as human, an important part of our lives.

The horizontal and vertical traffic inside the Lufthansa headquarter building is clearly and explicitly organized. Users and visitors can easily find out the directions in the building. The corridor of the central axis is connected with the office and the atrium, and the vertical traffic is organized through three sets of transparent glass elevators to direct people into each floor. Visitors can position themselves accurately through the transparent glass, whether they are in the elevator, the central axis corridor or the atrium. The crescent-shaped space of the axis corridor provided the visitors the possibility of direct communication.

4

5

7

6

项目解读：德国法兰克福汉莎航空公司总部办公楼
建 筑 师：英恩霍文建筑师事务所

CASE STUDY: LUFTHANSA AG HEADQUARTER BUILDING, FRANKFURT, GERMANY
ARCHITECTS: INGENHOVEN ARCHITECTS

LUFTHANSA AG HEADQUARTER BUILDING, FRANKFURT, GERMANY

汉莎航空公司总部办公楼项目，成功设计了一个零能源、零排放的中庭空间。在这一项目中中庭起到均衡温度、通风节能的作用，并且赋予了其新的意义——花园办公模式和风景区的感受。新中庭产生了新的功效，减轻了工作压力，成功改变了传统的办公形式，轻松、愉悦的办公环境直接提高了人们的办公效率。

大楼整个建筑由不同的类似手指的狭长单体和单体间的中庭花园构成。在第一期的建筑中提供了1650个工作位，并形成了标准化、单元式的建筑形式。总体建筑由29个狭长单体构成，在每层约15000㎡的建筑面积中拥有4500个工作位。

中轴线的走廊是建筑的交通枢纽，它连接着所有区域并且使人们轻松地辨别方向。弓形的天棚形成了一个奇特的空间。花园中庭还设计了自然的幕墙通风系统，并起到了阻挡噪声和辐射的作用。中庭花园的绿色植物来源于五大洲，形成了不同形式的花园风格，表达了汉莎航空公司"洲际连接"的含义。

The horizontal and vertical traffic inside the Lufthansa headquarter building is clearly and explicitly organized. Users and visitors can easily find out the directions in the building. The corridor of the central axis is connected with the office and the atrium, and the vertical traffic is organized through three sets of transparent glass elevators to direct people into each floor. Visitors can position themselves accurately through the transparent glass, whether they are in the elevator, the central axis corridor or the atrium. The crescent-shaped space of the axis corridor provided the visitors the possibility of direct communication.

The project of Lufthansa headquarter building successfully designed an atrium space with zero energy and zero emission. The atrium functions as an energy saving ventilator and regulates the temperature. And it is also endowed with a new meaning – the garden office mode and the scenic enjoyment.

The new atrium converts the conventional office space to a relaxed and pleasant office environment, reducing the working pressure and improving the working efficiency.

The entire building is composited of figure-shaped units and the garden atrium in between. The first phase of the building provides 1650 seats, formalizing a standardized, modular construction. The complete building has 20 fingers, and each floor covers about 15,000 square meters, providing 4500 seats. The corridor on the central axis is the traffic hub, connecting all 29 figures and indicating the direction. The arched ceiling forms a peculiar space, and the garden atrium is also designed with glass façade equipped with natural ventilation system, blocking out noise and radiation. The greenery from the five continents establishes various garden styles in the atrium, implying the intercontinental connection of Lufthansa.

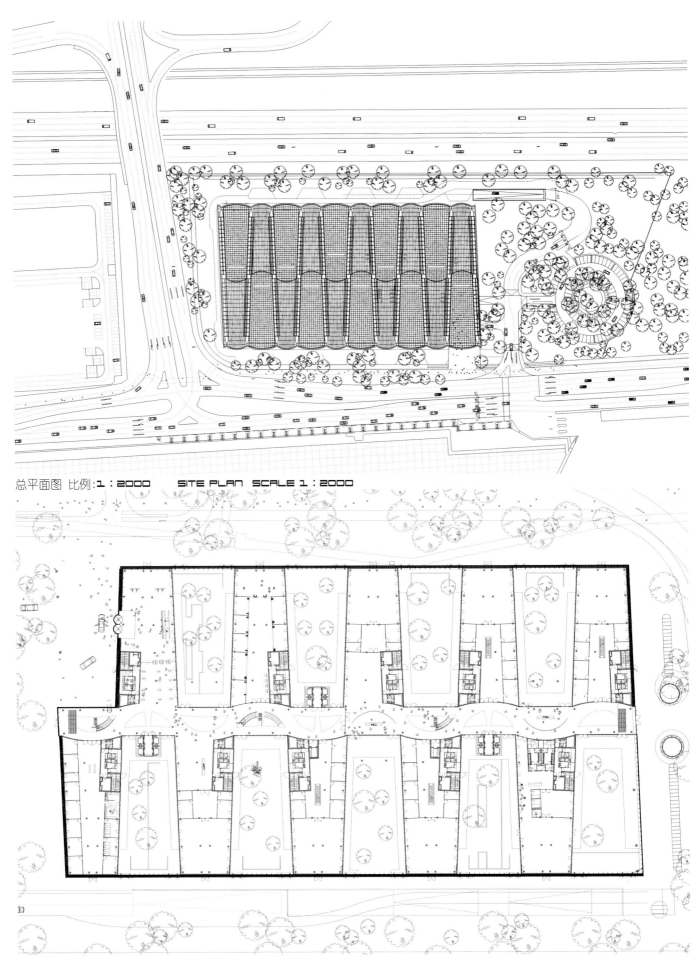

总平面图 比例:1:2000　　SITE PLAN SCALE 1:2000

一层平面图 比例:1:1100　　GROUND FLOOR SCALE 1:1100

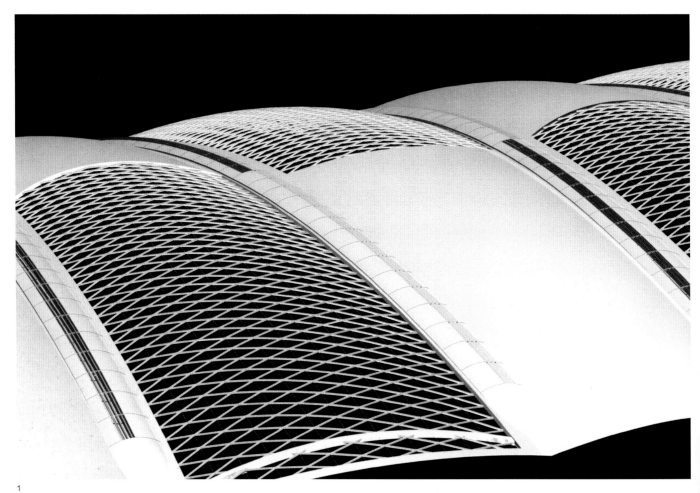

1

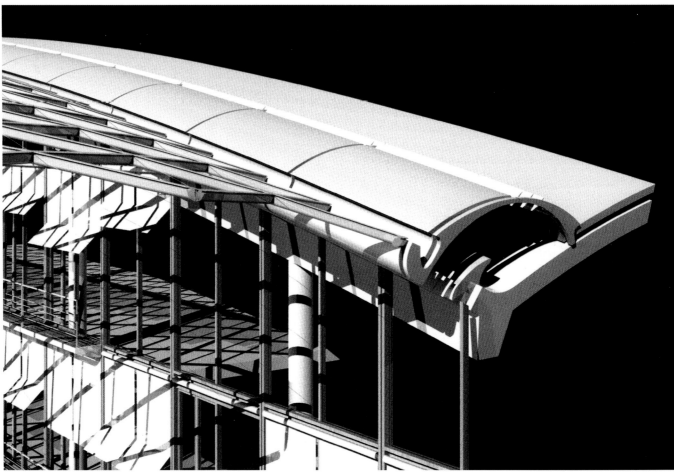

2

3

4

5

6

7

1. 在设计过程中,我们就对钢筋混凝土弓形屋顶和玻璃弓形屋顶的连接进行了结构的分析,多功能构件集雨水排放、排风、排烟为一体,被安装在两个弧形屋顶之间的交接处
2. 设计图片
3. 模型分析
4. 通过专业软件将灯光的分布进行模拟实验并进行优化,计算机模拟图片展示了未来空间的明亮度
5. 模型照片合成图
6. 建成照片
7. 建成照片

1. In the design process, we analyzed both the steel and the glass arched roof structures. This versatile component incorporates drainage, ventilation and smoke exhaustion, it is electronic driven, computer controlled to turn on/off, and installed at the joint of the two curved roofs. Through specialized software simulation, the light distribution is optimized, and the computer simulation picture shows the brightness of the future space.
2. Design drawing
3. Model analysis
4. By professional software distribution of the light simulation and optimization, computer simulation picture shows the brightness of the future space
5. Photoshoped model photos
6. Photo of the building
7. Photo of the building

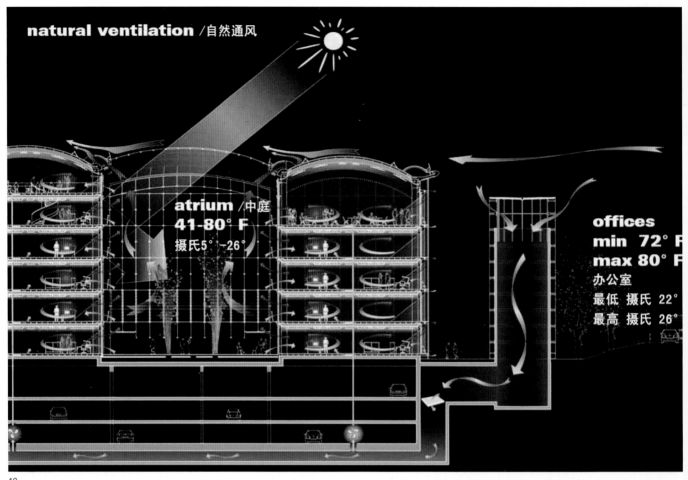

10

11

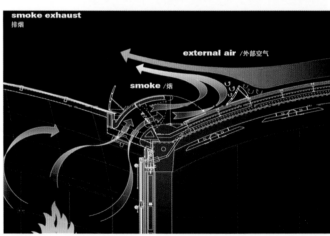

12

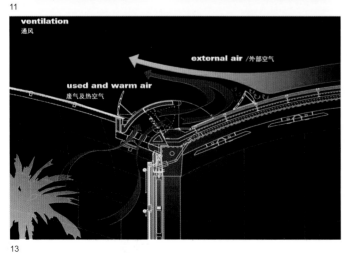

13

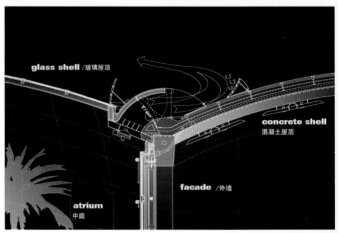

14

15

8. 亚琛空气动力研究所通过沙池1∶1模型实验,解释了通风口在各种状态下的排风状况,对空气动力进行了优化
9. 多功能排风排烟构件模型。
10. 自然通风系统图
11. 排水系统图
12. 排烟系统图
13. 排风系统图
14. 构造系统图
15. 效果图

8. Institute of Aerodynamics at Aachen University did a 1∶1 model experiment by using sand in a pool to explain the exhausting condition through exhaust vent under all circumstances hence to optimize the air force
9. Multifunctional exhaust component model
10. Natural ventilation
11. Drainage system
12. Smoke exhaust
13. Ventilation
14. Structural system
15. Renderings

12 体量和外表
VOLUME AND SURFACE

对建筑体量和外表的优化是现代建筑重要的、必需的工作程序，体量和外表影响了大楼的能源消耗，从而直接影响大楼的节能问题。

The optimization of building volume and surface is an important and necessary procedure of modern architecture design, which has decisive impact on the energy consumption of the building, hence directly affects the energy-saving design of the building.

12 体量和外表
VOLUME AND SURFACE

表面

表面的概念是材料形体表面的媒介。

"体量"一词来源于拉丁语"Volvere"。体量意为"转动、滚动"。

外表在自然界的优化

生物和植物在求生的过程中必须自身优化才得以生存，如同大象的皮肤或者鲸鱼的皮肤。为了防止太阳照射造成皮肤干燥，大象在进化过程中就尽可能减少了躯干外表的大小。

对比树叶和树干，树干庞大的体量有利于获得太阳的照射和雨水的收集，树叶则通过阳光照射来进行光合作用制造自身所需要的养分。

大楼体量的优化和大楼的外墙面

大楼外墙表面优化的目的是使用绿色材料，满足对能源节约、安全、生态的要求。

一个圆柱形的建筑和其他造型的建筑在面积大致相同的状况下，圆柱形建筑可以将太阳照射到外墙的面积减少到最低点。大楼的直径（包括办公空间的进深）应该是根据自然光的利用和通风的要求来决定的。

德国商业银行圆柱形的形体就满足了能源、室内温度、风压及经济的要求。

德国商业银行通过对平面的组织，使必要的功能空间，如井道、洗手间、茶室都集中于核心筒两侧。位于办公室和过道的隔断墙顶部也设置了高置窗，使过道通过高置窗来获得自然光线。大楼的直径为42m，它是根据节能的最佳体量而确定的。

1. 大象的体量和外表有其逻辑性
2. 树的体量和外表亦具逻辑性
3. 自然的外表 —— 甘肃张掖丹霞地貌
4. 北京国家大剧院的体量和外表无论时间（时代）、地点（城市规划）都是一个正确的逻辑决定　建筑师：保罗·安德鲁
5. 西班牙兰萨罗特博物馆　建筑师：英恩霍文建筑师事务所

1. The body size and the appearance of elephants have their logical reasons
2. Trees also have their logical reasons
3. Natural appearance - Gansu Zhangye Danxia landform
4. Regardless of time (Time) location (Urban Planning), the Chinese Theater in Beijing is a correct logical decision in respect of size and appearance. Architect: Paul Andreu
5. Musum Las Maretas Lanzarote. Architects: Ingenhoven Architects

4

VOLUME AND SURFACE

Surface

The concept of surface refers to the medium plane of a body of material touching another.

"Volume" comes from the Latin word "Volvere", the concept of rotating and scrolling.

Optimization of Surface in Nature

Living creatures must go through the process of self-optimization to be able to survive. For example, to prevent the dehydration from sun irradiation, the elephant scaled down its torso surface area as much as possible in evolution.

On the contrary, a bigger tree trunk is conducive to collect sun irradiation and rain falls, and larger leaves transfer more solar energy through photosynthesis.

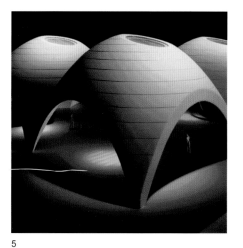

5

Optimization of Building Volume and Façade

The optimization of building's façade aimed to use Green building materials and to meet the requirements on energy conservation, ecological protection and safety.

A cylindrical building can minimize the surface area thus the solar irradiance, compared with building of other shape with similar footage. The diameter of the building (including the depth of office space) should be decided based on the use of natural light and ventilation requirements.

The cylindrical building of German Commerzbank successfully deals with energy consumption, indoor temperature control, wind pressure and economic cost.

The layout plan of German Commerzbank building placed the necessary functional space such as technical well, toilet and tea room to the both side of the core tube. The top of the partition wall between the office and the hallway is furnished with high-set windows, through which the natural light gets into the hallway. The building's diameter is 42m, optimized for efficient energy saving.

6

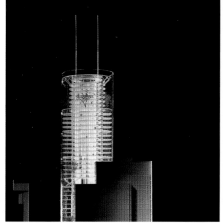

7

6. 德国法兰克福商业银行的平面图 建筑师：英恩霍文建筑师事务所
7. 德国法兰克福商业银行的模型 建筑师：英恩霍文建筑师事务所

6. Commerzbanks floor plan, Frankfurt Germany. Architects: Ingenhoven Architects
7. Commerzbanks model, Frankfurt Germany. Architects: Ingenhoven Architects

项目解读：德国埃森火车站
建 筑 师：英恩霍文建筑师事务所
CASE STUDY: RAILWAY STATION, ESSEN, GERMANY
ARCHITECTS: INGENHOVEN ARCHITECTS

RAILWAY STATION, ESSEN, GERMANY

埃森火车站位于城市的中心，它将城市分为两个部分。在埃森市政厅的规划构想中，希望通过新火车站的建设将城市的两部分有效连接。

建筑师在这一项目的设计中提出了一个巨大的弧形屋顶的造型构思，使大跨度的弧形屋顶将站台罩住，并成为埃森的"城市之门"，有效地连接了城市的两部分。

新的接待厅连接了埃森城市的商业街和商务区及位于另外一侧的维利·布兰特广场。在新火车站的设计方案中通过内部功能的更新和现代建筑的塑造给服务业、餐饮业、商业提供了更多的商业机会，并且成为现代的城市交通中心。

站台内的平面功能设计通过地上层和地下层的组织，将地铁与站台相连，使乘客能够在最短的时间内换乘其他交通工具，从而达到现代化火车站的高效要求。

埃森火车站的功能分区为：一层为接待大厅，二层为站台，地下一到三层为辅助商业区并与地铁连接。圆形的全景电梯可以自地下三层直接到接待大厅和站台。

一层的中心接待大厅设置了旅行社、等候区和咨询处，透明的玻璃电梯和扶梯连接了站台内的各个层面。接待大厅位于弧形外壳的下方，外壳结构的跨度达到90m，钢结构的框架承载了铝合金的外壳，在入口处设计了开放的屋顶，并通过大面积的玻璃面使乘客直接识别站台、火车。

玻璃外墙、地面灯光及具有反射功能的吊顶灯光系统，在夜晚给乘客提供了一个温和、明亮的室内环境，厅内的灯光无限向外扩散，构成了一幅时尚、现代又美轮美奂的景象。

The railway station of Essen is located in the center of the city, dividing the city into two parts. In the urban planning by Essen city hall, the construction of a new railway station should effectively connect the two parts of the city.

The architects proposed an idea of a huge arc roof, spanning over the railway platforms. It symbolized the "City Gate" of Essen to connect the two parts of the city.

The new reception hall connects the Essen commercial street and the business district on one side, and the Willy-Brand-Square on the other. The updated internal function and the new modern architecture attract more business opportunities for service, catering and commercial business, and make the railway station a modern transportation center of the city.

The subway and the railway platform are connected through multi-levels above- and under-ground, so that passengers can transfer to other transportations in the shortest time as to achieve the high efficiency of the modern railway station.

The first floor of Essen railway station is the reception hall, the second floor is the flat forms, and the first to the third floor underground is the shopping area connected to the subway. A cylindrical panoramic elevator carried visitors from the third floor underground directly to the reception hall and platforms.

The reception hall in 1st floor is occupied by travel agencies, waiting areas and information center, and the transparent glass elevators and escalators connect to all levels of the platforms. The reception hall is located right beneath the arc-shaped shell which has a span of 90m. The steel framework sustains the aluminum alloy shell, and the roof is designed open at the entrance.

The glass façade, the ground illuminating and the reflective ceiling lighting system provide passengers a mild, bright indoor environment at night. The light in the hall diffuses indefinitely outward, constituting a stylish, modern, and beautiful scene.

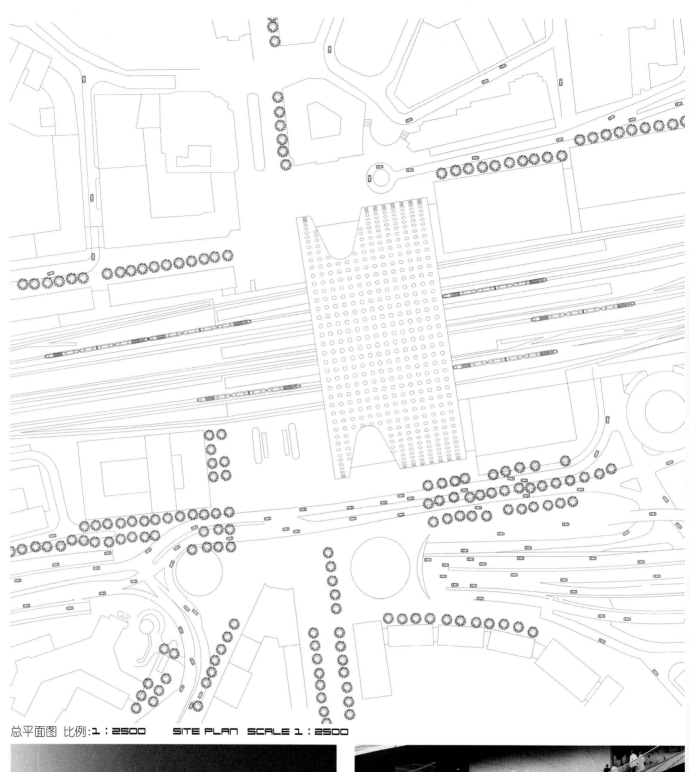

总平面图 比例：1：2500　　SITE PLAN SCALE 1：2500

项目解读：兰州盛达金城广场城市综合体
建 筑 师：德雅视界建筑师事务所 + 主创建筑师卢卡斯
CASE STUDY: SHENGDA JINCHENG PLAZA CITY COMPLEX, LANZHOU
ARCHITECT: IDEAL ARCHITECTS + LUCAS ZIMNY

SHENGDA JINCHENG PLAZA CITY COMPLEX, LANZHOU

兰州盛达金城广场城市综合体是由两座相同形状的塔楼构成，两座塔楼拥有一个共同的裙房。一座为办公楼，另一座为五星级酒店及酒店式公寓，酒店位于塔楼的底层，酒店式公寓则布置在塔楼的高层。

两座塔楼高约199m，其中办公楼共有40层，每层的层高是3.9m。五星级酒店及公寓共有43层，每层的层高是3.6m。裙房共有8层，1层的层高为6m，2~7层的层高为4.8m，裙房8层的功能分区为：1~6层为商业，7~8层为五星级酒店。

天水中路和南昌路的十字路口成为基地最重要的交叉路口，项目就坐落在这一十字路口上。两座塔楼形成了"城市门"的印象。两座塔楼之间最近的距离为20m。这一建筑主要展示了建筑向城市开放的概念，并为市民提供了宽敞的商业入口和城市广场。位于基地东南侧的天水中路这一醒目位置的是该建筑的前广场，这里为小轿车、出租车、大巴车提供了足够的停车空间。办公的主入口位于南昌路上，公寓的主入口布置在基地南侧，酒店的入口则布置在东南侧。所有裙房的外墙立面都可以提供给商业作为橱窗使用，商业的货用平台位于基地的西侧。

该项目使用了现代的建筑语言，外墙的构思来源于兰州的黄河及水车博览园，主要体现了流水像瀑布一样的外墙主题。三角形的叶片被布置在外墙上，并均匀地延伸和转化，行人会感受到一个变化的、具有动感的外墙。如果行人改变观察角度，将会观察到另一个不同的外观。如果你是开车路过，你将会有黄河水自上而下流动的感觉。

"城市门"的概念重点表现在两座塔楼之间的"玻璃形体"上。玻璃形体是简单的全玻璃的形体，它使观察者感受到两座塔楼之间的交流，形体的线条是自一个开始到另一个结束。酒店的"玻璃形体"处布置了套房，在顶部安排了完美的瞭望区。办公楼的"玻璃形体"处则安排了总裁办公室。"玻璃形体"使用了低辐射中空玻璃，外墙的叶片可以使用石材或铝合金材料，此处的设计使用了铝合金型材。裙房的功能共分为两个部分：商业和酒店服务功能。裙房的商业有1个小的中庭，中庭内设计了观光电梯。2个自动扶梯分别设计在临近东北入口和西南入口处，裙房的1~6层均为商业用房。

一层酒店区域设计了餐厅、接待、商务中心、钢琴酒吧、公寓大堂及咖啡吧。酒店的服务功能设计在裙房的7层和8层，7层设计了各种会议厅、KTV娱乐设施、SPA、健身区及泳池。8层设计了各种餐厅、包房、早餐厅、厨房，东北侧设计了大宴会厅及辅助设施。酒店标准层为24间客房，酒店共设计了360间客房，另外还提供了312间公寓。办公标准层面积约为1576㎡，办公空间可布置单间办公模式及大空间办公模式，可以根据租用的要求进行弹性设计。

Shengda Jincheng Plaza City Complex in Lanzhou city is a building complex constituted by two towers of the same shape and an adjoining subsidiary building. One of the two towers is foreseen as an office building, the other a five-star hotel and apartments. The lower levels of the tower are destined for the hotel, while the higher levels for the apartments.

Both towers are about 199 meters high, with 3.9 m height between floors the office building will accommodate 40 floors; and the five-star hotel and apartments accommodates 43 floors, with floor-to-ceiling height of 3.6 meters. The adjoined building is consist of 8 floors, with the height between floors of 6 m for the 1st floor and 4.8 m for the 2nd to 7th floors. The functions are distributed as: 1st to 6th floors for commercial, 7th and 8th floors for hotel services.

The crossway of Tianshui Zhong Road and Nanchang Road is the most important intersection for the lot. As the project is located right on this crossway, the uniform towers are placed parallel to those two streets and formed the "City Gate". The closest distance between the two towers is 20 meters. This building complex shows the open concept of the city and provides the public a spacious entrance of shopping center as well as a city square. Located on the southeast side along Tianshui Zhong Road, the front square of Jincheng this building occupies a striking location, provides enough space for cars, taxis, and buses (to pass through and load passengers). The main entrance of the office building is located on Nanchang Road, while the main entrance of the apartments is arranged in the south and the hotel entrance in the southeast. The exterior façade of the adjoining subsidiary building can be used as window show for commercial purpose, and the freight platform is arranged on the west side of the lot.

The project uses a modern architectural language: the idea of external walls comes from the Yellow River and the Lanzhou Waterwheel Exposition, as the theme of waterfall is symbolically represented by the façade. The leaves (blades) of triangle shape are arranged on the façade, gradually extending and transforming, making the pedestrians feel a dynamic changing exterior wall. When changing their viewing angle, the pedestrians will get a different look; and by driving, people will feel the water flowing from top to bottom as it was from the Yellow River.

The concept of "City gate" is represented by the "glass bodies" between the two towers, which depicts the meaning of exchange by the smooth lines from one to the other. Hotel suits are placed inside of the "glass body" to get a perfect view and in the other tower the CEO Office (President Office) correspondently. LowE glass is to be used for the "glass body". The blades on the exterior wall could be made of natural stone or aluminum alloy, and the design chooses to use the aluminum alloy profiles. Functionally, the annex building is planned to have both commercial and hotel service purposes. There is a small atrium inside the commercial section of the annex building, within which a sightseeing elevator is designated. Escalators are destined on the northeast and southwest axis respectively, as from 1st to 6th floors the annex is intended to be commercial.

For the hotel section, the reception lobby, restaurant, business center, piano bar, coffee bar and the lobby for hotel apartments are all destined on the 1st floor. The hotel service facilities and departments are located in the annex building on the 7th and 8th floor: meeting rooms, KTV, SPA, Fitness & Pool are planned for the 7th floor; restaurants of all cuisines, private dining rooms, breakfast area, kitchen as well as the Grand Ballroom and other auxiliary facilities on the 8th floor. The hotel standard layer 24 rooms, the hotel designed a total of 360 rooms. With the area of 1576 square meters, the hotel's standard floor has 24 guest rooms. In total, the hotel provided 360 guest rooms and 312 apartments. Area of the standard office floor is approximately 1576 square meters. Office space may be arranged in the single-office model and/or open-space office model, which can be designed flexibly according to the leasing requirements.

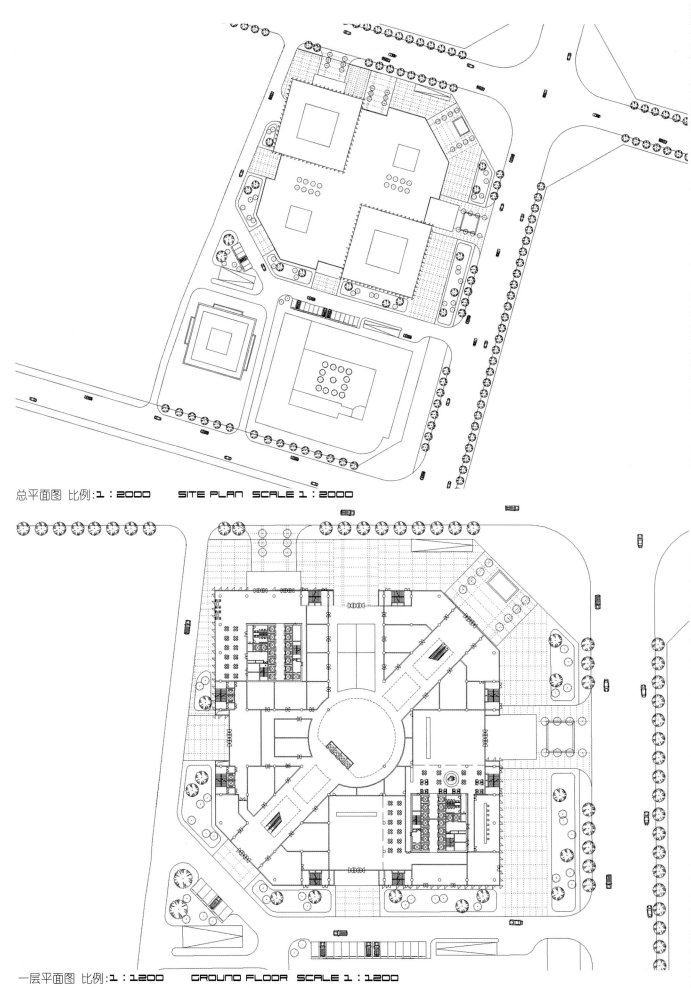

总平面图 比例：1：2000　　SITE PLAN SCALE 1：2000

一层平面图 比例：1：1200　　GROUND FLOOR SCALE 1：1200

13 效率
EFFICIENCY

使用少量而达到多量是我们对效率的核心要求。少量的能源需求而达到众多的使用功效，这一概念应该成为建筑师在建筑设计过程中的重要任务，而"高效"应该成为新时代建筑的重要评判标准。

THE CORE INTENT OF EFFICIENCY IS TO USE LESS TO ACHIEVE MORE. USING LESS ENERGY TO ACHIEVE MORE FUNCTIONS SHOULD BE AN IMPORTANT TASK OF THE ARCHITECTURE DESIGN. THUS THE EFFICACY SHOULD BECOME A KEY CONCEPT OF THE NEW ERA CONSTRUCTION.

13 效率
EFFICIENCY

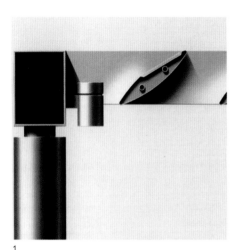

1
1. 莱茵集团总部大楼的太阳能光电接受板
2. 杜塞尔多夫大学教学楼，高效的节能设计使所有数据达到德国绿色建筑标准　建筑师：英恩霍文建筑师事务所
3. 汉莎航空公司总部多功能排风排烟构件模型
4. 莱茵集团总部大楼通风构件
5. 莱茵集团总部大楼外墙节点

   1- 固定件　　　　6- 阻热
   2- 单元件　　　　7- 眩光保护帘
   3- 通风、出风口　8- 采暖设备
   4- 遮阳帘　　　　9- 金属板
   5- 内推拉窗

1. RWE headquarters building solar panels
2. Oeconomicum, University Düsseldorf. High efficient, energy-saving design, all data match with the green building standards of Germany (DGNB). Architects: Ingenhoven Architecture
3. Lufthansa AG headquarters multifunctional exhaust smoke component model
4. RWE headquarters building ventilation component
5. RWE headquarters building façade detail

   1- mount　　　　　6- heat isolation
   2- unit component　7- anti-glare curtain
   3- ventilation outlet　8- heating equipment
   4- sunscreen curtain　9- stand-on metal plate
   5- inner sliding window

"效率"一词来源于拉丁语"efficere"，是"影响"的意思。
"效率"的概念是指影响、能力或在经济学中有效的方法。

必要性

所有事物应该衡量其自身的必要性，没有必要的建筑构件在大楼运行的过程中肯定会很少使用，优秀建筑的构件基础是来源于事物的必要性。

建筑师在规划建筑的过程中应该区别本质和非本质、必要和非必要来作为优化设计的基础。

共同的必要性将帮助我们在规划过程中区分有意义和无意义的构思及其建议。

高效的节能建筑构件

大楼的建筑构件、连续构件及来自工厂加工完成的构件有其特殊的要求和目的。这些构件和完整的系统应用的目的在于，它必要的、统一的效率和经济因素。

莱茵集团总部大楼拥有一个高效的、节能的外墙构件，我们称它为"鱼嘴构件"，也是第一代模拟的通风构件，它包括了所有必要的功能需要，如外层玻璃、内层玻璃、气候保护的空气廊、通风口、采暖设备及必要的外墙清洗功能的平台（空气廊的金属板可以让人站立清洗玻璃）。

这一完整的、封闭的单元外墙构件是在工厂内生产和加工的，并且考虑运输的安全性，最终在现场安装完成。

数字双层幕墙

汉沙航空公司总部办公楼是基于莱茵集团大楼的模拟系统发展的数字双层幕墙系统，这一系统可根据天气的变化控制通风量，从而形成舒适的通风环境。阻隔噪声的玻璃还具备了防辐射的功能。此外，在办公室外安装的金属遮阳帘，高效地阻止了热量进入室内空间，保证了室内温度的均衡。

EFFICIENCY

2

The concept of efficiency comes from the Latin word "efficere", meaning the "impact". "Efficiency" refers to the impact, the capacity or the effective method in economics.

Necessity

Everything should be weighted by its necessity. Un-necessary components used in building are rarely used in operation. Therefore, in an efficient architectural design the building components used are based on their necessity.

In the architectural planning process, a distinction should be made between essentiality and non-essentiality, necessity and non-necessity as the foundation for optimal design.

Common necessities will help us to distinguish between meaningful and meaningless concepts and suggestions in the planning process.

3

High-efficient Energy-saving Building Components

Building components, connecting components and factory manufactured modules have their own special requirements and applications. Yet they all share common goals: necessity, energy efficiency and economic benefits.

The façade unit of RWE headquarter building is a highly efficient, energy-saving façade component which is called the "fish mouth". It is the first generation of analog ventilation module that includes all function components such as outer and inner layer of glass, air gallery for climate protection, ventilator, heating equipment and façade cleaning platform (the stand-on metal plate in the air gallery).

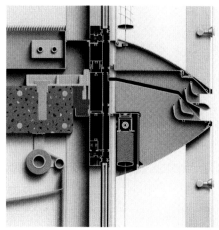

4

The complete, enclosed unit facade module is custom-built and assembled in the factory. The final installation is done on the construction site, considered the safety of transportation.

Digital Double Layer Curtain Wall

The digital double facade of the Lufthansa Airlines headquarter office building is developed from the analog facade system of the RWE building. The noise/radiation proof glass and the metal sunscreen curtain efficiently prevent the heat penetration, ensure an equalized indoor temperature.

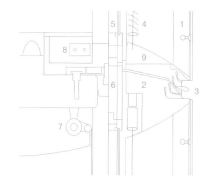

5

项目解读：海口西海岸假日酒店
建 筑 师：德雅视界建筑师事务所+中国建筑技术集团有限公司
室内设计：文格空间设计顾问(深圳)公司

CASE STUDY: HAIKOU WESTCOAST HOLIDAY INN, HAIKOU
ARCHITECTS: IDEAL ARCHITECTS + CHINA BUILDING TECHNIQUE GROUP CO.,LTD
INTERIOR DESIGN: SHENZHEN WENGE INTERIOR DESIGN CO.,LTD

HAIKOU WESTCOAST HOLIDAY INN, HAIKOU

海口西海岸假日酒店的设计主题是创造一个现代的海岸酒店，使每个客房都能看到大海。建筑由三层酒店裙房和十二层客房构成。酒店位于海口西海岸，基地隔海口景观大道——滨海大道——与北侧海面相望。酒店的主楼形体为蝴蝶状双塔造型，裙房为由斜线和直线构成的超现代风格的形体。

酒店的平面设计包括了所有服务功能，如餐厅、会议中心、健身房、SPA等。酒店的服务设施为客人提供了星级服务。酒店前侧设计了规模较大的入口广场，广场中心设计了具雕塑感的音乐喷泉，这一喷泉是景观轴的始点，并成为酒店的迎宾入口。

当客人进入酒店时，第一时间将来到透明、宽大的酒店大堂，体验现代酒店的时尚和魅力。酒店大堂位于基地的景观轴上。透过大堂可以向南观看到景观轴的绿色景观。酒店大堂被塑造成一个贯通三层的中庭空间，自然的阳光可通过玻璃墙面进入大堂中庭。大堂中心区设计了接待、咨询、休息等设施。大堂南侧为大堂吧，客人可坐在大堂吧内观赏室外的水景和绿化。酒店的二层是相对私密的功能区，包括餐厅、会议室及SPA。

酒店三层为宴会厅及健身房。四层是酒店的接待大堂，当客人乘坐专用电梯自一层到达四层时可以看到透明的接待大堂，近距离感受大海。接待大堂两侧为商务酒廊及早餐厅。四至十五层为酒店客房，每个房间都可以看到大海。

The design theme of Haikou Westcoast Holiday Inn is a modern coastal hotel featuring ocean view from every guest room. The building is composed of 12-storey hotel guest rooms and 3-storey annex.

The hotel is located in the west coast of Haikou, facing north to the sea across Jingguan/Binhai Blvd. The hotel's main building is butterfly-shaped twin towers, and the annex building has a post-modern shape framed by slash and straight lines.

The layout design of the hotel covers all service functions, such as restaurants, conference center, gym and SPA, providing the high quality of star-rating service for hotel guests. In the front of the hotel is the larger entrance plaza. At the center of the plaza, a sculptural music sprinkler fountain stands as the starting point of the landscape axis to connect the overall landscape, marking the welcoming entrance of the hotel.

As soon as they make the entrance, guests will feel the transparency and spaciousness of the hotel lobby, and experience the fashion and charm of modern hotel. The hotel lobby is located right on the landscape axis, through which guests can enjoy the green landscape to the south. It is a 3-story atrium featuring natural lighting through the glass façade. Natural light fell through the glass façade into the lobby atrium which is a 3-stroy loft space. Reception, information, rest area and other facilities take the place in the center of the lobby. On the south side is the lobby bar, where guests can relax and enjoy the outdoor water features and greenery. The second floor of the hotel is a relatively private function area, including restaurants, conference rooms and a Spa.

The 3rd floor is furnished with banquet halls and a gym. The 4th floor is the reception lobby for the hotel guest rooms. Ascending from the 1st floor to the 4th floor in the dedicated elevator guests will be impressed by the transparent reception lobby and enjoy the ocean view close by. The business lounge and the breakfast restaurant are located on both sides along the reception lobby. From 4th to the 15th floor are all hotel guest rooms, each with its own spectacular ocean view.

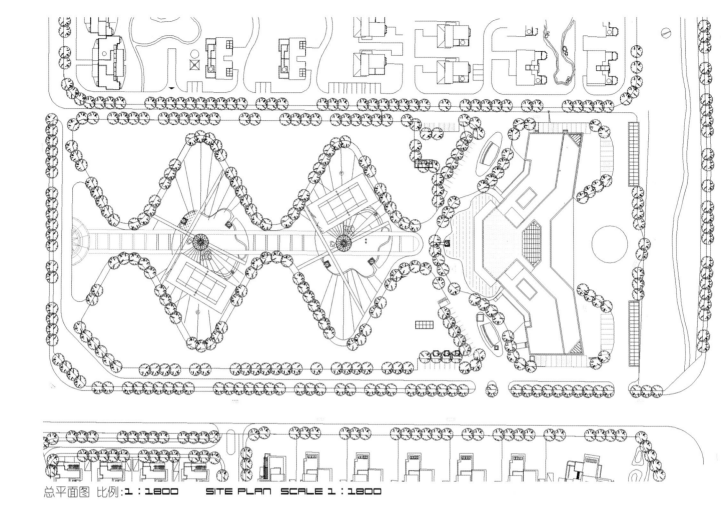

总平面图 比例:1:1800　　SITE PLAN SCALE 1:1800

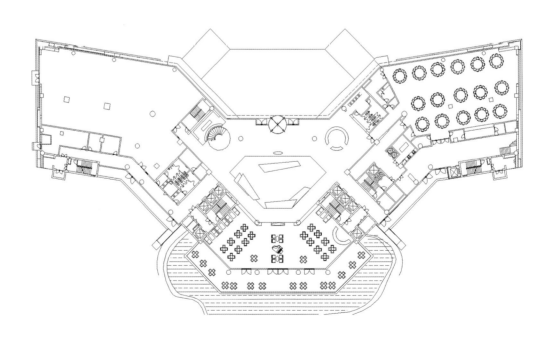

一层平面图 比例:1:800　　GROUND FLOOR SCALE 1:800

14 个性
PERSONALITY

建筑的独特性、感染力、认同感可以成为具有进化意义的成果，也可以成为建筑形象的内在动力及城市标志。优秀的建筑表现了城市鲜明的个性，人们也通过建筑的个性了解到城市的文化与历史。

THE IDENTITY, IMPRESSIVENESS AND ABILITY TO CONVINCE OF A BUILDING WILL HAVE EVOLUTIONARY SIGNIFICANCE. THEY MAY SERVE AS THE UNDERLYING POWER OF THE BUILDING IMAGE AND A CITY LANDMARK. EXCELLENT ARCHITECTURE TELLS DEFINITELY WHAT A CITY IS, WHILE ON THE OTHER HAND, PEOPLE ARE ALLOWED TO BE EXPOSED TO THE CULTURE AND HISTORY OF THAT CITY.

14 个性
PERSONALITY

1

一座充满艺术魅力的城市，必定是具有鲜明的文化个性的城市、必定是建筑文化积累最丰富的城市。

个性

"个性"一词最初来源于拉丁语"Personalitatem"，是指个人的自我特征。

建筑个性

建筑通过建筑师的塑造，特别是通过节点和构造的设计、材料及技术的应用，使其在与城市规划或景观的关系的平衡中，确定了它的体量、形式和个性。

重庆自然博物馆的设计，其外观体量表现了极少主义的内涵，简化的建筑形体与自然景观的融合是这一建筑的鲜明个性。

合肥招商银行大厦通过四个转换延伸的组合体，表现了其运动、转换和融合的概念和鲜明的建筑个性，它明确的体量和其运动的造型将在城市中迅速显现并成为标志。建筑的个性是城市某种知觉的反映，知觉是文化的显现，是生命存在的理由和标志。建筑，责无旁贷地担负着这个使命。

1. 德国法兰克福MAX高层综合体　建筑师：英恩霍文建筑师事务所
2. 海口海岸公馆公寓楼——错位的个性魅力
3. 科隆波恩机场的塑造和组织使旅客首先感受到其外观形式和机场的空间，然后认知大跨度的空间结构和绿化，最终产生了强烈的感受　建筑师：英恩霍文建筑师事务所
4. 合肥招商银行大厦——四个长方形体组合
5. 德国汉堡新博览会展厅　建筑师：英恩霍文建筑师事务所

6. 汉诺威世博场馆——圆形组合　建筑师：英恩霍文建筑师事务所
7. 西宁雅啦索酒店——切角组合　建筑师：德雅视界建筑师事务所

1. Highrise complex MAX Frankfurt am Main Germany. Architect: Ingenhoven Architects
2. Haikou Mansion apartment building- Dislocation charismatic personality. Architects: Ideal Architects
3. The shape and organization of the Cologne-Bonn Airport make the passengers first acknowledging its appearance, form and space, and then realizing the large-span structure and the green area, ultimately rising the strong awareness of recognition. Architect: Ingenhoven Architects
4. China Merchants Bank- Four rectangular combination
5. New Trade Fair Hamburg. Architects: Ingenhoven Architects
6. EXPO-Plaza Honnover. Architects: Ingenhoven Architects
7. The Xi Nining Yalasuo Hotel - Cutaway combination. Architects: Ideal Architects

2

PERSONALITY

3

A city full of artistic charm must be a city with distinct cultural personality and rich in accumulation of architectural culture.

Personality
The term of personality, originated from the Latin "personalitatem", refers to the self-concept of individual.

Architectural Personality
A building's volume, form and personality is determined by the architects' design, stressing the composition of detail and structure as well as the application of materials and technology to reach its balance in urban planning or landscaping.

4

In the design of Chongqing Museum of Natural History, the museum's exterior/appearance and volume represented the connotation of minimalism, and the simplified architectural form integrated with the natural landscape is the distinct character of this building.

China Merchants Bank Building in Hefei is composed of four converting and extending elements, representing the concept of movement, transformation and integration. Its distinctive architectural personality will make the building promptly apparent and become the city flag. The architectural personality is a reflection of a certain perception of the city, and the perception is the manifestation of the culture, and the reasons and signs of life's existence. Architecture takes on this duty-bound mission indefinitely.

5

7

6

项目解读：合肥MV广场
建 筑 师：郭小平
CASE STUDY: MV PLAZA, HEFEI
ARCHITECTS: XIAOPING GUO

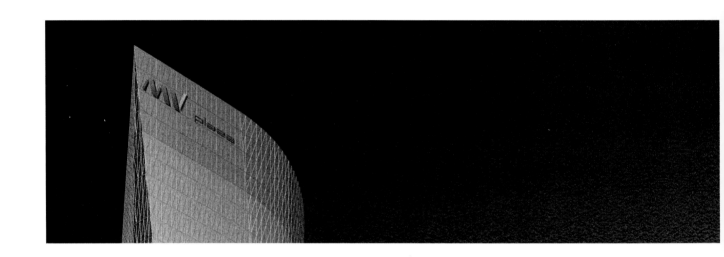

MV PLAZA, HEFEI

MV广场的单体是由一座180m和一座120m的高楼构成，它的单体是一个自下而上由圆形变为方形和一个自上而下由方形变为圆形的单曲建筑造型。在视觉上它由方变圆，由圆变方；在数学上，它的形态是由底层到顶层旋转了180°角的造型。

在纷繁芜杂的现象背后，总有某种规律在控制着。正如牛顿用"F= ma"来概括力学现象一样，我们试图用"建筑形体＝基本形＋形式逻辑"来体现形体生成的逻辑性。方与圆这两个原始的基本形，自从被人类发现以来，就得到了广泛的运用，这一方案试图探索这两个基本形的内在逻辑，平面取方形和它的内切圆的组合，将方与圆两个基本形的精华糅合为一体。

方与圆从表面上看差别极大，似乎是两个极端。而本方案采用了渐变的逻辑，将圆变成方的所有逻辑过程展现出来，让人们看到这两个基本形之间的整个演变过程。

融合与冲突

这种"基本形＋形式逻辑"所得到的结果是令人惊喜的，它体现出了方与圆的融合与冲突。它们的融合体现在：在得到的形体中，母线是垂直的，这给结构的实现带来了方便；它们的冲突则在渐变的过程中得到两条倾斜的斜线。这两条斜线使形体表现出倾斜的姿态，给人们以强大的视觉冲击力。

组合

对两个倾斜形体的镜像组合，平衡了单个形体单方向倾斜的力量，使建筑组合体得到了力的平衡。

成果评价

形式生成和组合的结果是能充分满足建造的要求和体现功能的性格的。所得到的单个形体母线是垂直的，结构可以采用常规的筒中筒的结构形式。而其表面是由两个平行四边形、两个三角形和四个四分之一圆锥面围合而成的。在幕墙设计上能通过简单的三角玻璃来实现。在对两个单体的组合中，呈现出了一个"V"的形象，是英文"victry"的首字母，代表着胜利，同时也代表了现代办公楼的气质和品性。其形式简洁而又富于变化，不失为一个能够代表新时代合肥的建筑形象。

The MV Plaza is constituted by two high-rise buildings, with the height of 180m and 120m respectively. One tower transfers from cylinder to cubic bottom up, so does the other top down. The formation visually represents the conversion between round and square, and mathematically rotates 180 degrees upside down.

Beneath chaos there rules the law. As Newton discovered "F=ma" in classic mechanics, we tried to use "architecture formation = basic form + formal logic" to describe the logic in shape formation. Square and Circle, the project attempted to explore the internal logic of these two basic shapes. From the top view, the circle inscribes the square, representing the kneading of the two basic shapes as one.

The shapes of Square and Circle vary considerably towards two extremes. The design used the logic of the gradient, unfolding the process of circle transforming into square, thus people can see the transition between these two basic forms.

Fusion and Conflict

"Basic form + formal logic" approach brings a pleasant surprise, reflecting the fusion and conflict of Square and Circle. The fusion defines the vertical structure of the building, which is convenient to implement. The conflict outlines two inclined curves in the transition, creating a leaning posture with an overwhelming visual impact.

Combination

Combining two mirrored inclined bodies equalizes the leaning force of either element, obtaining the balance of forces of the whole structure.

Result Evaluation

The formation and combination of the shapes fully meet the requirement of construction, and reflect the character of the building's function. The vertical mainline of the body shape allows the conventional structure of tube in tube. The surface of the body is composed of two parallelograms, two triangles and four quarters of conical surface. The design of glass façade is realized by simple triangular pieces of glass. The combination of the two building units forms the image of a "V", symbolizing "victory" and representing the quality and character of the modern office building. A simple and neat form, yet full of change, that would be a representative architectural image of the new era of Hefei.

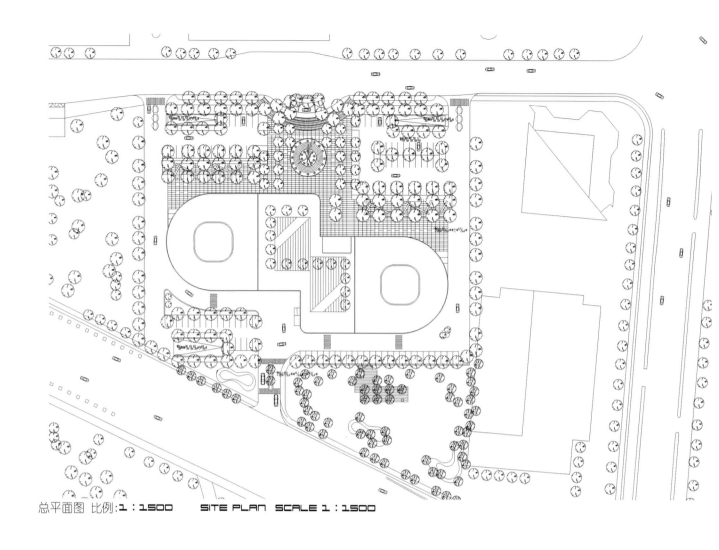

总平面图 比例：1：1500　　SITE PLAN SCALE 1：1500

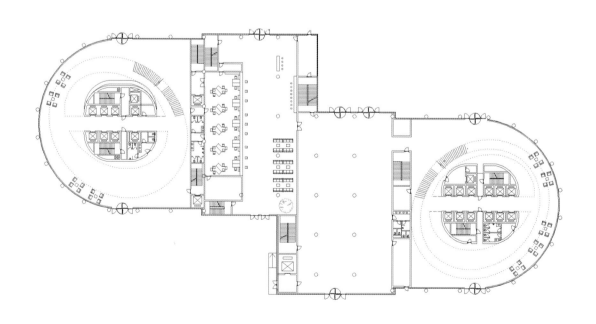

一层平面图 比例：1：800　　GROUND FLOOR SCALE 1：800

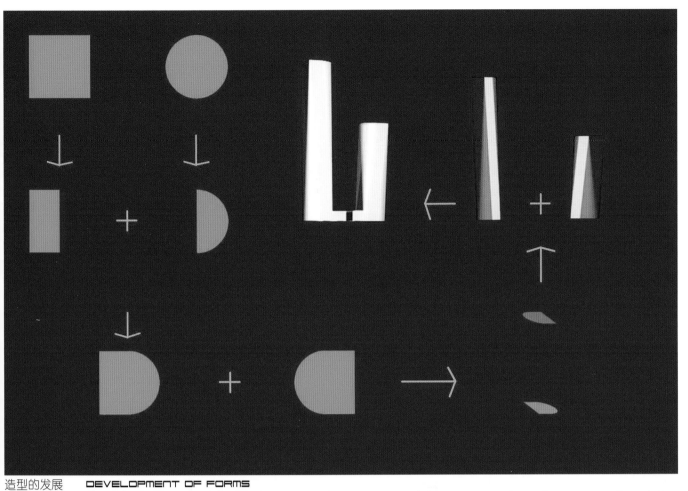

造型的发展　DEVELOPMENT OF FORMS

平面的蜕变过程　PLAN TRANSFORMING PROCESS

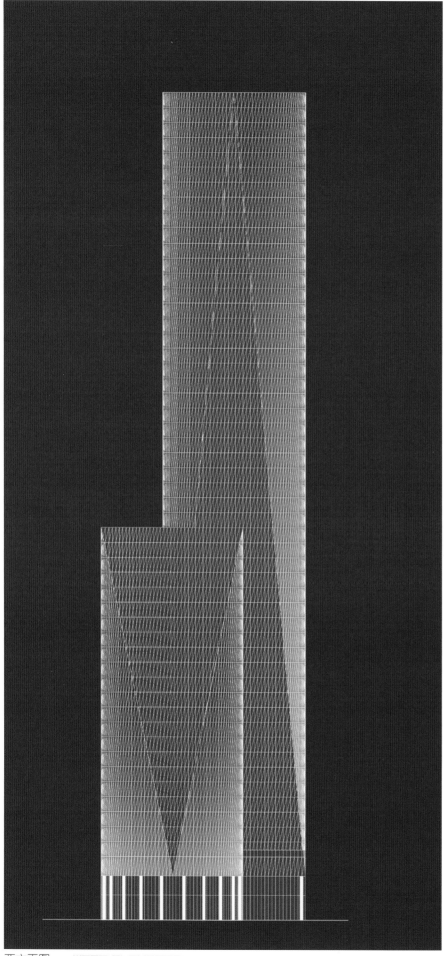

西立面图　WEST ELEVATION

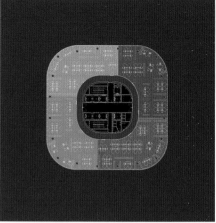

13层平面图　13 TH FLOOR

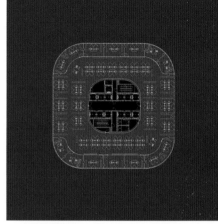

27层平面图　27 TH FLOOR

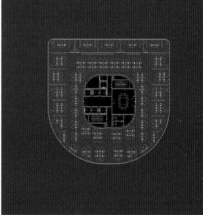

50层平面图　50 TH FLOOR

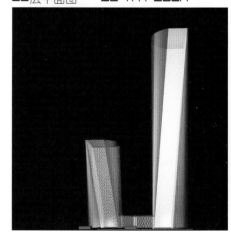

项目解读：兰州盛达金城广场城市综合体
建 筑 师：德雅视界建筑师事务所
CASE STUDY: SHENGDA JINCHENG PLAZA CITY COMPLEX, LANZHOU
ARCHITECTS: IDEAL ARCHITECTS

基地位于兰州市中心区——盘旋路附近,东临天水中路,南侧是旅游局大楼,经天水路直接同兰州火车站相连,北侧经雁滩后是G30高速公路的出口。

基地呈长方形,整体建筑群由两座塔楼和一座7层商业裙房构成,2栋塔楼分别布置在基地的西北侧和东南侧,商业裙房布置在基地的东北侧,基地的西南侧则形成了庭园。北侧塔楼A为度假酒店和酒店式公寓,共30层,高158m;南侧塔楼B为五星级酒店和办公室,共62层,高229m;东北侧为商业裙房,共7层,高36m。

简洁、有力、时尚的设计语言是我们对新建筑的设计理念和追求。由两座塔楼和裙房构成的建筑综合体直接表现了建筑概念,塔楼的造型是由两个三角形构成的方形,方形还做了30°角切面,其中一个三角形的延伸并构成了建筑的造型,这个经切面的三角形同另一个三角形构成了高差,错落有致,既变化又统一。

裙房的功能共分为两个部分:商业和酒店服务功能。裙房的商业有一个小的中庭,中庭内设计了观光电梯。自动扶梯分别设计在东北和西南轴线上,裙房的1~6层均为商业。一层酒店区域设计了餐厅、接待室、商务中心、钢琴酒吧、公寓大堂及咖啡吧。酒店的服务功能设置在塔楼2~7及裙房的7层,7层设计了各种会议厅、大宴会厅及辅助设施。

该建筑是新的人文景观,它具有强烈的视觉表现力和冲击力,它联系着一种公众精神、聚合力和城市所承载的首创精神。

The lot is located in the downtown area of Lanzhou. Close to Panxuan Road, this rectangular lot faces Tianshui Zhong Road to the east, is adjacent to the Tourism Office Building in the south, connected with Lanzhou Railway Station through Tian-Shui Road, and linked to G30 highway on the north side by Yantan.

The building complex consists of two tower buildings and a 7-story adjoining commercial building. The two towers were arranged on the northwest and the southeast side, while the commercial annex on the northeast, making the southwest side an inner courtyard. Tower A on the northwest side is designed as a resort hotel including hotel apartments with a total of 30 floors and a height of 158 meters; Tower B is planned as a five-star hotel and high-end office building with a total of 62 floors and height of 229 meters; and the adjoining commercial building with a total of 7 floors is 36 meters high.

Simplicity and strength, the stylish design represents our idea and pursuit of new architecture. The building complex of two towers and the annex demonstrated our architectural concept: the towers formed a square constituted by two triangles, and that square through a 30° section made one of the triangle stretching to the architectural configuration which in confront with the other triangle framed the height difference, posing a changing yet unified progress.

Functionally, the annex building is planned to have both commercial and hotel service purposes. There is a small atrium inside the commercial section of the annex building, within which a sightseeing elevator is designated. Escalators are destined on the northeast and southwest axis respectively, as from 1st to 6th floors the annex is intended to be commercial. For the hotel section, the reception lobby, restaurant, business center, piano bar, coffee bar and the lobby for hotel apartments are all destined on the 1st floor. The hotel service facilities and departments are located in the tower building from 2nd to 7th floor as well as on the 7th floor of the annex building, e.g. multiple meeting rooms, large ballroom and other auxiliary facilities.

Architecture is new cultural landscape; it has a substantial visual expression and impact, associated with the public vitality, cohesion, and the pioneering spirit hosted by the city.

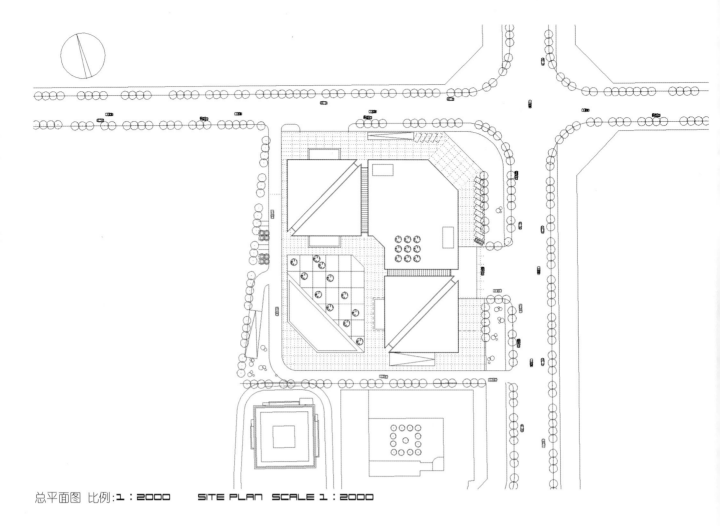

总平面图 比例：1：2000　　SITE PLAN SCALE 1：2000

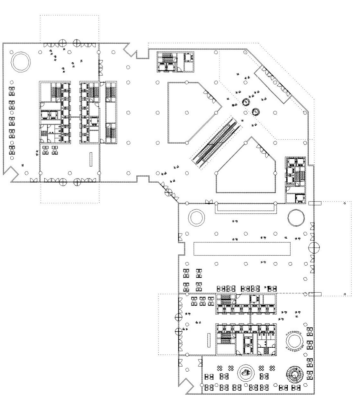

一层平面图 比例：1：1100　　GROUND FLOOR SCALE 1：1100

15 美
BEAUTY

美和时尚是一个建筑是否合理的评判依据。合理的建筑大多是美的，而不合理的建筑往往是丑陋的。

Beauty and fashion in architecture are the right decision to make. Most right buildings are beautiful, while wrong buildings are often ugly.

15 美
BEAUTY

1
1. 鲸鱼是生活在太平洋里的哺乳动物，它最长可达33m，最重可达136t。它是从陆地的哺乳动物发展到海洋的哺乳动物
2. 古典的美——巴塞罗那圣家族教堂顶棚 建筑师：安东尼·高迪
3. 我们达到一定的高度和要求，同时又展现了进化之前的成绩：美和正确
4. 自然美的形式
5. 木材

1. The whale is a mammal living in the ocean, measured up to 33 meters and weighted up to 135 tons. It evolved from land mammals to the marine mammals
2. Classic beauty - Sagrada Familian Church Ceiling. Architect: Antonio Gaud
3. We reach a certain height and requirements, at the same time show the evolution of the previous results: the United States and the correct
4. Beauty in natural form
5. Timber

　　美是一个必要的逻辑成果，是逻辑、真实、效率、简单、效应、少量、使用、放松的概念的延伸——如同芭蕾舞演员的精湛表演，如同一个画家完成作品后的喜悦——从压力、矛盾、投入、辛勤的工作经过长时间的发展成形。使用和激情的关系、生活的轨迹、材料转换的效果都将成为美的部分。

真实
　　美的概念来源于古希腊语"aisthesis"——真实的意思。如果看上去像木材，那它应该就是木材，塑料制成的产品应该有其不同于木材的制造工艺。它们质地不同属性也就不同。我们应该信任事物的内在品质，信任空间、心理和物理的真实。因此，建筑师应该在建筑设计的过程中保留材料的真实性。真实是正确建筑的前提，也是必需的。保留事物的真实使我们设计与生活变得容易和轻松。

绿锈
　　气候造成的氧化、化学和空气的作用在金属面的表现，我们称它为"绿锈"。教堂屋顶的绿锈表现其历史的真实性。

木材
　　木材、石材及骨材同属于人类最古老的材料。
　　自人类历史以来，木材作为主要的建筑材料而被使用，同时，家具、武器、工具、船舶、马车都使用木材作为基本材料。

老化、使用、痕迹
　　年龄是生物和植物生活持续发展的结果，衰老是自然的、生物的进化过程，是化学物理的现象。时间的演化改变了事物的结构，而这种变化取决于时间的推进。建筑通过材料的质地展示了美、技术和经济的内涵，它们通过自然技术的影响和使用改变了原材料的质地。
　　文化的转换关系，它包括了道德、精神及人类形而上学的观点，同时决定了完整的美学。

Beauty is essentially a logical outcome, the extension of the concepts such as logic, truth, efficiency, simplicity, effect, minimalism, usage and relaxation – like the superb (exquisite) performance of the ballet dancer or the joyful feeling of a painter finishing

BEAUTY

hisart work. Beauty comes from a long way of pressure, contradiction, devotion and hard work. The practice vs. passion, the trajectory of life, the effect of material conversion, all these will become part of beauty.

Authentic

The concept of authenticity is derived from the ancient Greek word "aisthesis", meaning the truth. If it looks like wood, it should be wood indeed. A plastic product is different from a wood product in their manufacture process, as well as their texture and characteristics. We must trust the intrinsic quality of matters, and believe the true feeling of space, psychology and physics; and as architects we should preserve the material's authenticity in the design process. Authenticity is the premise and necessity of a right building. Leaving matters to their authenticity makes our design as well our lives much easier.

Patina

Patina is the oxidation caused by whether conditions, a chemical reaction of air with metal surface. The patina on a church roof presents the authenticity of its history.

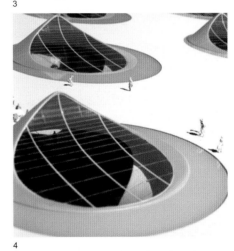

Timber/wood

Wood, stone and bone are among the oldest tools people used.
Wood has been used as main building material all through the human history. Furniture, weapons, tools, ships and carriages use wood as the basic material as well.

Age, Usage, and Trace

Age records the continuous development of biological lives, thus aging is a natural process out of biological evolution, a phenomenon of chemistry and physics. The impetus of time alters the structure of matter. Buildings showcase the aesthetic, technical and economic connotations through the texture of the material, and the texture of building materials changes due to the influence of nature and the application of technology.

The culture transformation in perspective of morality, spiritual pursuit, and metaphysics are also dedicated to the aesthetics.

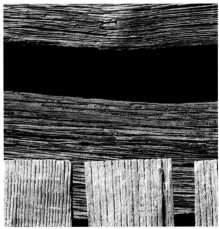

项目解读：德国汉堡空客A380组装大厅
建 筑 师：英恩霍文建筑师事务所
CASE STUDY: AIRBUS A380 ASSEMBLY HALL, HAMBURG, GERMANY
ARCHITECTS: INGENHOVEN ARCHITECTS

AIRBUS A380 ASSEMBLY HALL, HAMBURG, GERMANY

欧洲航天航空公司EADS主要由欧盟成员国德、法、西、英四国构建而成，总部设在法国的图卢兹。巨型飞机A380是该公司的重要机型，由各国分担发动机、机翼、机身的生产，最终的组装选择在德国的汉堡完成。为此，空客公司举行了组装大厅的方案招标，英恩霍文·欧文迪克事务所被邀参加投标。

基地位于一个生态自然保护区内，东临易北河，这就意味着这一组装大厅的塑造必须结合生态景观，在项目设计的开始，建筑师就意识到长750m的大厅使用传统的结构方式是非常困难的，而是必须使用现代的结构形式构筑大厅，体量尽可能地弱化，以便保留风景区的生态景观。为了使方案顺利进行，建筑师们选择与结构工程师索贝克合作，索贝克构思了具备伸缩性、无柱的大跨度屋顶系统，并且匹配了多种技术可能。组装大厅的功能主要由封闭的完成区和传统组装区构成，这两部分必须没有任何阻碍，玻璃推拉门为10m×21m，组装大厅的人们可以通过玻璃门观望到自然保护区的景观。结构设计师们选择了非对称的承重体系。位于易北河的一面没有支撑柱，也没有高度的限制，通过100m的悬挑结构可以使外墙面同大厅高度相等。主支撑柱承载整体屋顶，由钢拉力件交叉固定。根据几何学的原理，结构工程师们优化了钢材的使用量。

屋顶结构设计为优制钢材，并具备防腐蚀功效。自然的光线通过屋顶的折射面进入大厅，使大厅在通常状况下无需人工照明就可获得足够的光线。屋顶还具备排风和排烟的功能，自然的空气通过百叶窗进入大厅，进行自然通风。组装大厅的封闭完成区和系统组装区没有设计机械的通风系统，只有油漆车间设置了机械通风和空调技术系统。大厅的屋顶由不锈钢材料焊接构成，材料表面为亚光，具备无方向的反射效果，这样可以将自然光和灯光反射在大厅中，优化了大厅的灯光系统。空客A380的组装大厅成为这一自然保护区的一部分，让人们真实地感受到这一生态建筑物的魅力。

空客大厅的设计为建筑师和结构工程师的紧密合作提供了范例，因为建筑形式的塑造同结构紧密相连。支撑臂的悬挑覆盖了整个大厅。支撑臂由钢构件通过标准件组合构成，模式化的标准件使分期建造成为可能，在工厂里加工的支撑臂可以通过船运到目的地，并在升降机上焊接，最后进行升降安装。支撑臂在没有其他支撑柱的状况下悬挑100m，覆盖了整个大厅，使大厅内空客380飞机的组装不受任何影响。

AIRBUS A380 ASSEMBLY HALL, HAMBURG, GERMANY

The European Aerospace Company (EADS) is constituted by the main EU member countries: Germany, France, Spain and Great Britain, headquartered in Toulouse of France. The giant A380 aircraft is an important model made by the company. The production of engine, wings and fuselage is distributed among different countries, while the final assembly is completed in Hamburg, Germany. Airbus called for a tender for the design of the Assembly Hall, at which the Ingenhoven & Overdick firm of architects was invited to participate.

The construction site is located in an ecological nature reserve area, east of the Elbe Lake, which commands that the architectural configuration of the building must be concordant with the ecological landscaping. At the beginning phase of the design, architects were aware of the difficulty of using the traditional structure to support the 750 m long hall. Therefore, the modern structure must be applied to reduce the building volume as much as possible, so that the ecological landscape of the scenic area can be preserved. The structural engineer, Sobek (Wemerv.Sobek), conceived a stretchable column-free large span roof system and matched a variety of technical possibilities. The function of the assembly hall is mainly carried out in the enclosed finishing zone and the traditional assembly area, and no obstruction is allowed in these working zones. The sliding glass doors measure 10m × 21m, through which displays the scene of the nature preserve through which displays the scene of the nature preserve. The structure engineers selected the asymmetric bearing system – the lake side is free of supporting columns and height limits, running the facade as high as the hall through the 100m cantilever structure. The main sporting columns carry the whole roof, cross-fixed by steel tensions. The usage of steel is optimized according to the principles of geometry. The roof structure is made of premium steel with anti-corrosion treatment. Natural light enters the hall through the refracting surface of the roof, eliminating the needs of artificial lighting in the usual situation. The roof is also equipped with air/smoke extractors, and the blinds allow the natural air inflow. The mechanical ventilation system is not installed in the enclosed finishing zone and the traditional assembly area, only the paint shop is designed to equip with mechanical ventilation and air conditioning system. The roof of the hall is constructed by the stainless steel welding, and the matt material surface allows diffusion reflection to improve the lighting in the hall. The assembly hall of Airbus A380 has become part of the Nature Preserve, and people are impressed with the charm of this ecological building.

The design project of the assembly hall for Airbus A380 provided an example of close cooperation between architects and structural engineers, as the configuration of building is closely linked to the structural feasibility. The supporting arms are composed of steel components in standard modules. The use of standard modules enabled the staging construction; the supporting arms manufactured at the factory were transported to the destination by ship, welded and installed using lifting equipment. Steel reinforcing has been optimized during the design process of construction. The supporting arms cantilevered 100m over the entire hall, free of any other footholds otherwise may disturb the assembly of Airbus A380.

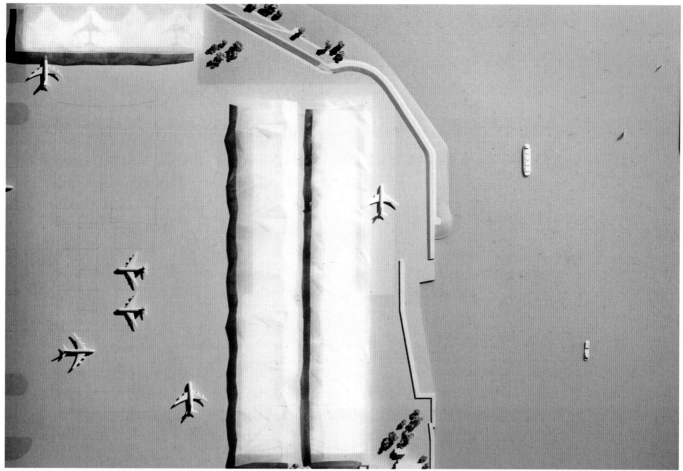
平面模型　PLANE MODEL

结构模型　STRUCTURAL MODEL

结构模型　STRUCTURAL MODEL

结构模型：钢结构桁架　STRUCTURAL MODEL: STEEL STRUCTURE TRUSS

16 未来建筑——零能源建筑
FUTURE ARCHITECTURE - ZERO-ENERGY BUILDING

我们设想——未来建筑是在没有其他辅助的能源帮助下运行的。建筑本身成为一个收集器，它能将风能、光能、地热等再生能源、新能源收集、转换、使用，再收集、再转换、再使用，循环往复，以至永远。

　　我们设想——通过建筑师与结构工程师、技术顾问、外墙专家、能源工程师、建筑物理学家等众多专业的技术人员协同配合，开发革命性的节能技术并应用于建筑中，创造零能源、零排放的未来建筑。

LET'S ENVISION THAT FUTURE BUILDINGS WILL OPERATE WITHOUT ANY AUXILIARY ENERGY. THE BUILDING ITSELF IS A COLLECTOR HARVESTING THE ENERGIES AND NATURAL RESOURCES, INCL. WIND, SOLAR, GEOTHERMAL HEAT AND OTHER RENEWABLE AND NEW ENERGIES. THE CIRCULATION OF COLLECTION, CONVERSION, APPLICATION AND RE-COLLECTION, RE-CONVERSION AND RE-USE CAN GO OVER AND OVER AGAIN.

FOR THE FUTURE BUILDING WE ASSUME THAT ARCHITECTS AND STRUCTURAL ENGINEERS, TECHNICAL ADVISERS, THE FAÇADE EXPERTS, ENERGY ENGINEERS, BUILDING PHYSICISTS AND MANY OTHER PROFESSIONALS AND TECHNICAL PERSONNEL WILL WORK TOGETHER CLOSELY. THEY AIM TO DEVELOP REVOLUTIONARY ENERGY-SAVING TECHNOLOGIES TO APPLY TO BUILDING CONSTRUCTION, AND TO CREATE ZERO-ENERGY, ZERO-EMISSION FUTURE ARCHITECTURE.

16 未来建筑——零能源建筑
FUTURE ARCHITECTURE - ZERO-ENERGY BUILDING

"未来"一词源于拉丁语"futures"。
零能源、零排放建筑是未来建筑的前提。

零能源城市

未来城市的能源系统将通过大面积的太阳能、风能、地热等自然能源构成能源供给框架。先进的节能建筑将减少能源消耗，构成城市建筑的节能框架；靠各种再生能源运行的电力车及自行车取代汽车的城市交通节能框架；城市的分类垃圾将通过现代化的化工技术手段重新加工使用，构成城市生态循环的框架……这些都将成为可持续城市发展的能源基础，并构成未来零能源城市的基础。

零能源建筑

实现零能源建筑是一个漫长的历程，我们应该从现在开始就尝试和研究，新能源的发展和应用，给我们提供了良好的能源基础，如风力发电、太阳能、地热等。随着技术的不断进步，将出现新形式的能源，它们可以作为零能源建筑的供给基础，同时建筑节能技术的发展和应用，为帮助实现零能源建筑提供了良好的条件，如高效节能幕墙、高效保温门窗、保温建材、高效供暖、节能设备、数字大楼控制系统等。新能源的使用和节能技术的应用，将构筑零能源、零排放的未来建筑。如果一个建筑物能达到60%的节能可能，那么另外40%的能源将需要通过新能源的供给获得，这样就可能会实现了零能源、零排放建筑的梦想。

1

1. 零能源办公楼 建筑师：英恩霍文建筑师事务所
2. 零能源建筑设计——金昌城市综合体 建筑师：德雅视界建筑师事务所
3. 零能源住宅——德国杜塞尔多夫住宅 建筑师：英恩霍文建筑师事务所
4. 同上
5. 零能源构想规划——德国Hammfeld Neuss 建筑师：英恩霍文建筑师事务所
6. 未来零能源建筑 图片：DETAIL Zeitschrift fuer Architektur Serie 2004, 7/8
7. Werner Sobek零能源住宅 R 128 图片：DETAIL Zeitschrift fuer Architektur Serie 2003, 10

1. Zero-energy office building. Architects: Ingenhoven Architects
2. Zero energy building-Jinchang city complex. Architects: Ideal Architects
3. Zero Energy House - Dusseldorf, Germany. Architects: Ingenhoven Architects
4. Ditto
5. Zero energy planning conception Hammfeld Neuss.Architects: Ingenhoven Architects
6. Future Zero Energy Building. Picture: DETAIL Zeitschrift fuer Architektur Serie 2004, 7/8
7. Werner Sobek zero energy house. Picture:DETAIL Zeitschrift fuer Architektur Serie 2003, 10

2

3
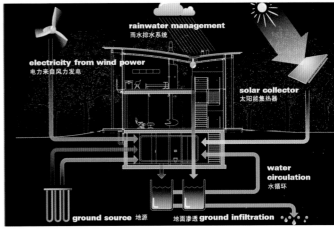
4

Derived from the Latin "futures", the concept of future is the indefinite period of time yet to come.
Zero-energy and zero-emission are the premise of future architecture.

Zero-Energy City

The future energy system of a city will be supported by a supply framework of natural energies such as: solar, wind, and geothermal heat. The state-of-the-art energy-efficient buildings will reduce energy consumption and thereby constitute the energy saving framework of urban building; electrical cars using renewable energy and bicycles will replace the traditional automobiles and therefore build up the energy-saving framework of urban transportation. The sorted urban garbage will be recycled through the processing of modern chemical industry and thereby contributes to the framework of urban biological circulation. All these will be the foundation of sustainable energy resources for urban development, and constitute the basis of the future zero-energy city.

Zero-Energy Building

The realization of zero-energy building is a long journey that we should try and study from now on. The development and application of new energies – such as wind power, solar, geothermal heat and etc. – has provided us a good energy foundation. As technology continues to progress, the emergence of new forms of energy will expand the supply basis for zero-energy buildings. At the same time, the development and application of energy-saving technologies provide good conditions to help realize the zero-energy building, e.g.: energy efficient façade, efficient insulation windows and doors, insulation materials, efficient heating, energy-saving devices, digital building control systems. The use of new energy and the application of energy saving technology will construct a zero-energy, zero-emission building of future: reducing energy consumption by 60% and new energy to supply the rest 40% will make the dream come true.

5

6

7

项目解读：德国斯图加特21世纪斯图加特火车站
建 筑 师：英恩霍文建筑师事务所
CASE STUDY: 21ST CENTURY TRAIN STATION, STUTTGART, GERMANY
ARCHITECTS: INGENHOVEN ARCHITECTS

21ST CENTURY TRAIN STATION, STUTTGART, GERMANY

背景

在国际著名结构工程师费赖·奥托配合下设计的斯图加特火车站以其开创性的空间构想在国际招标中获得一等奖，德国建筑专业杂志《房屋》评论：这一革命性的空间创造是继慕尼黑奥林匹克体育场之后最优美的建筑作品。斯图加特火车站是德国"21世纪斯图加特工程"中的一项，工程总投资约45亿欧元（约370亿人民币），其宗旨是：城市通过铁路设施的现代化获取利润，并要求轨道设施在城市开发中得以完善，从而提供便捷、高速的交通服务，以适应现代人的生活方式和节奏。占地面积约100hm²、站台长约400m（其中200m延伸到城市的中心绿地——宫殿花园）的斯图加特火车站位于老城的边沿及老火车站北侧。我们建筑师的构想是：保持宫殿花园的完整性，让轨道从地面消失，为城市创造一个绿色环境的未来空间。于是，一个童话般的构想产生了："我们做了一个火车站在地下，加上一个顶，几个洞，几个点。"

创作过程

建筑的本质在于它具有个性的发展过程，"Work in Progress"这个源于英语的词汇意指：为实现最初的目标所进行的一系列分析、推敲及完善工作。蛹吐丝成茧，最后变成蝴蝶的现象给我们这样的启示：一个自然而精美的形态总是在经过一系列的演变之后得以形成的。70年前由建筑师博拉茨设计的老火车站反映了当时的建筑风格和生活状况，21世纪的今天我们尝试了一个新的突破，自然形态与功能的完美组合。我们寻求一种从内容到形式都有别于常规的新火车站形式——轨道设施移至地下约6m处，以保留一个完整的宫殿花园。为了这个不同寻常的设想，我们进行了各种实验。通过工作模型、电脑模型，从城市关系、空间形成、能源节约到技术可行性及造价分析，进行了一系列反复的推敲与比较，其论证结果是：这不是一个幻想，它可以成为现实。

在此，城市建筑、生态与技术得到充分的糅合，这已不仅仅是一座建筑，更是一件艺术品。老火车站的塔体及设于四个方向的新火车站的玻璃壳入口，成为新火车站及城市的新标志。富有韵律的系列"光眼"构成了一幅梦幻般的画面，并作为老城与新城的连接链。这个流畅而开敞的空间连续而协调，从任何角度观看，这一空间都充满令人想象的魅力。

21ST CENTURY TRAIN STATION, STUTTGART, GERMANY

Background

In collaboration with Frei Otto, the internationally renowned structural engineer, the Stuttgart railway station won First Prize in an international tender for its innovative space vision. The German professional architecture magazine "Housing" (HAUSER) commented: this revolutionary space creation is the most beautiful architectural work following the Munich Olympic Stadium.

Stuttgart railway station is one of the "Stuttgart 21st Century Projects", which is budgeted a total investment of 4.5 billion Euros. The project aims to make the city more prosperous by the modernization of the railway facilities. The railway facilities should be improved along with the city development, providing a convenient and high speed transportation service to adapt to the fast-paced modern life style. Covering an area of ca. 100m, with the platform about 400m long (200m of which extends to the city's center green space – the palace garden), the railway station is located at the edge of the Old City and in the north of the old train station named "Bonatz Bau". To maintain the integrity of the palace garden, the architects decided to remove the track from the surface to create a future green space for the city. Thus, here comes the idea of fairy tale: "a railway station underground, plus a top, a few holes, a few points."

The Creative Process

The essence of architecture roots at its characteristic developing process. "Work in Progress" – this expression in English refers to a series of analysis, deliberation and optimization conducted to reach the initial goal. From a chrysalis to a cocoon, and a cocoon to a butterfly, the metamorphosis indicates the formation of the natural beauty through a series of evolution. The old train station (Bonatz-Bau) designed by the architect Bonatz 70 years ago represented the architectural style and living condition of that time. Today in the 21st century we tried a new breakthrough – a perfect combination of the natural form and the designed functionality. A railway station, innovative from content to form, is envisioned, moving tracks ca. six meters below ground to preserve the entire palace garden. To realize this unconventional idea, we conducted a variety of experiments using the working models and the computer models, from city relationship, space formation, energy saving to technical feasibility and cost analysis. A series of scrutiny and comparison led to one conclusion: it can become a reality instead of a fantasy.

In this case, city and architecture, ecology and technology are mixed together thoroughly and perfectly to create not just a building, but a piece of art. The glass shell entrance of the new train station facing all four directions, along with the old train station tower become the city's new logo. A series of full rhythmic "light eyes" constitutes a fantastic picture and functions as the connection between the Old Town and New Town. The smooth open space uses internal and external connections in a continuous space network; viewed from any angle, this space is filled with Imaginary charisma.

1. 总平面图
2. 卫星照片 ——宫殿花园
3. 效果图
4. 航拍照片 ——火车站及宫殿花园

1. Site plan
2. Satellite photos - palace gardens
3. Renderings
4. Aerial photographs - the train station and the palace gardens

1 总平面图 比例：1：3000　SITE PLAN SCALE 1：3000

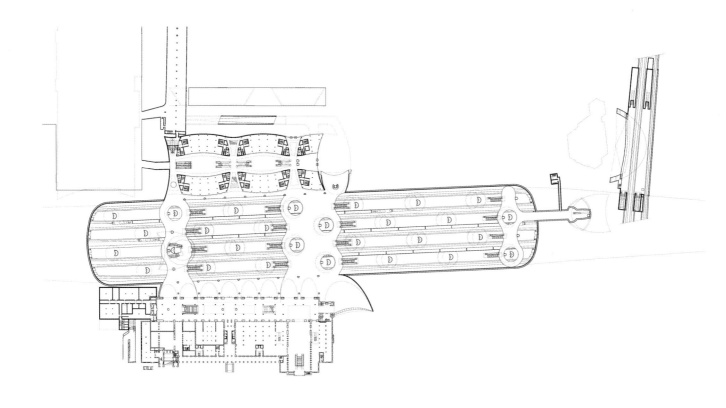

5 地下一层平面图 比例：1：4000　　1ST BASEMENT LEVEL　SCALE 1：4000

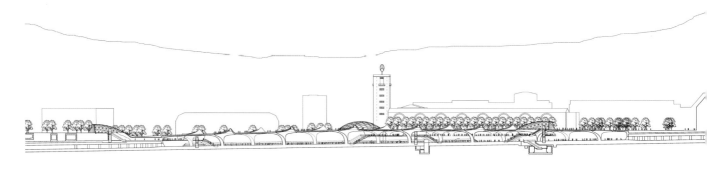

6 剖面图 比例：1：4000　　SECTION　SCALE 1：4000

5. 地下一层平面图
6. 剖面图
7. 模型照片
8. 德国ICE 第三代高速列车
9. 宫殿花园
10. 蝴蝶现象
11. 模型
12. 模型——"光眼"及入口
13. 3D模型——"光眼"

5. 1st Basement
6. Section
7. Model Photo
8. Third generation of the German ICE high-speed train
9. The palace gardens
10. Butterfly phenomenon
11. Model
12. model "Light eye" and entrance
13. 3D-model-"Light eye"

7

8

9

10

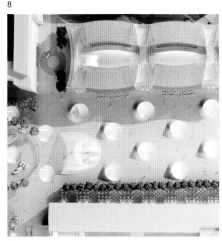

11

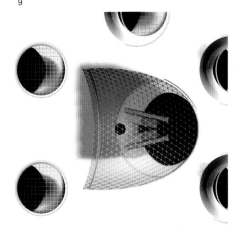

12

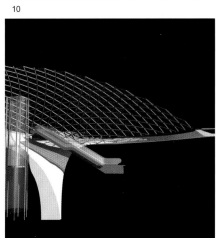

13

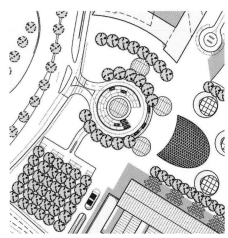
14 西入口平面图　WEST ENTRY FLOOR PLAN

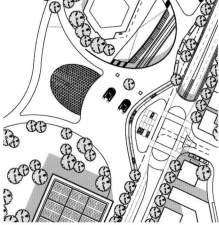
15 东入口平面图　EAST ENTRY FLOOR PLAN

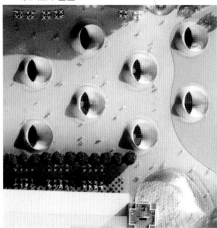
16 1:30工作模型　1:30 WORKING MODEL

18 1:30工作模型　1:30 WORKING MODEL

17 1:30工作模型　1:30 WORKING MODEL

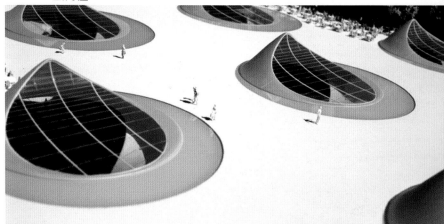
19 1:30工作模型　1:30 WORKING MODEL

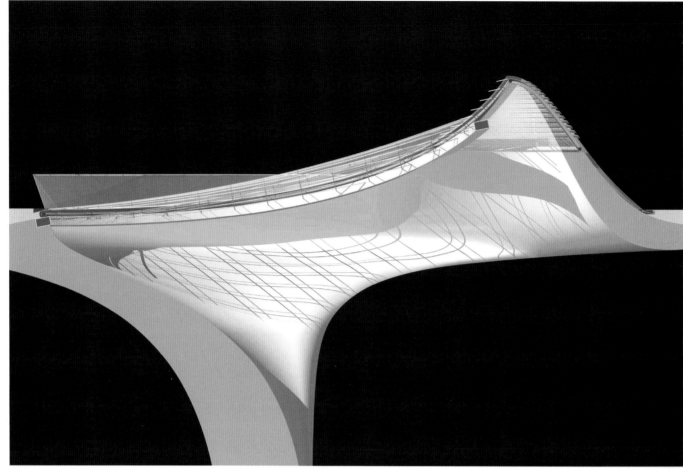

20

21

22

造型与结构

1963年弗赖·奥托通过肥皂薄膜进行了最小面积的受力实验，结果显示：在没有产生张力的情况下，要形成一个单力支撑的薄膜，必须通过一个孔的组合体才可实现，那就是"光眼"。在火车站的结构构思中，首先想到的便是钢网结构，这一构想要求"光眼"有一个精确的高度——如同肥皂模型所呈现的，钢网应与混凝土结合，并形成封闭的骨架——这样就构成了一个屋顶基础。结构计算的结果是：拉力荷载的悬挂屋顶代替压力荷载的水泥拱顶。这一形状通过模型进行实验，并在悬挂拉力的状况下凝固成形，然后旋转180°角即成为现在的形态——膜结构"光眼"。

20. "光眼" 3D 模型
21~22. 1963年弗赖·奥托通过肥皂薄膜进行了最小面积的受力实验
23~27. 弗赖·奥托的结构实验模型
28. 英国哈波尔德结构事务所的计算机模型图
29. "光眼" 模型
30~31. 受力分析
20. 3D-model- "Light eye"
21~22. In 1963, Frei Otto made an experiment to investigate the force distribution of the minimum area through soap membrane
23~27. The structure of an experimental model of structural engineer Frei Otto
28. Computer model of the structure of Happold (UK)
29. Model - "Light eye"
30~31. Stress Analysis

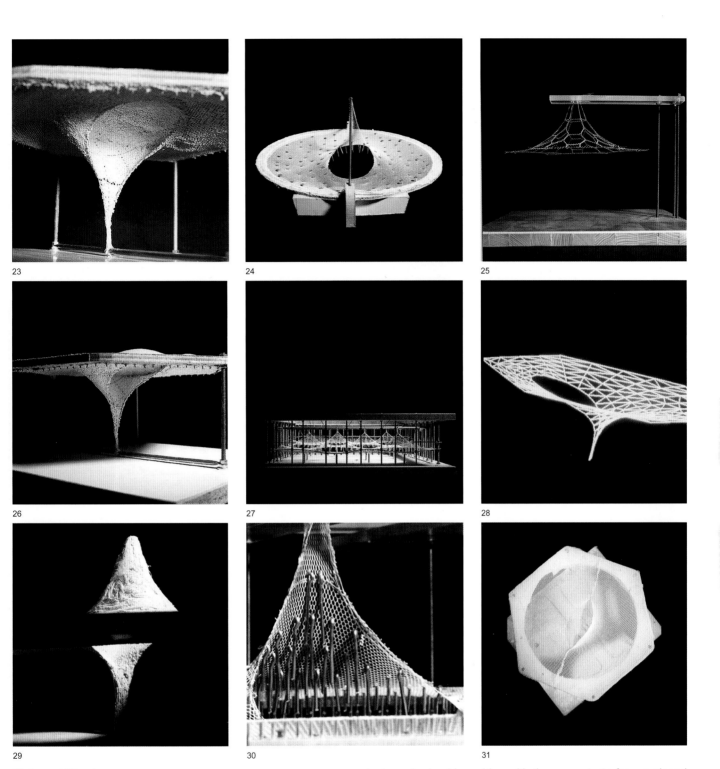

Style and Structure

In 1963, Frei Otto made an experiment to investigate the force distribution of the minimum area through soap membrane.

It is revealed that no tension was applied, a thin membrane supported by a single force can only form through the composite of holes call "light eye". The steel mesh structure was the first considered idea when it came to the structural design of the train station. This vision requested a precise height of the "light eye". As the model of soap membrane presented, the steel mesh should combine with the concrete to form a closed framework as the base of the roof. The results of structure calculation preferred the tension loaded suspension roof over the pressure loaded cement dome. This pattern was tested by the model experiments, solidified under the suspension tension, and then rotated 180 ° to get the current shape – the "light eye" membrane structure. This ideal structure shape is formed at a continuous and suitable edge condition.

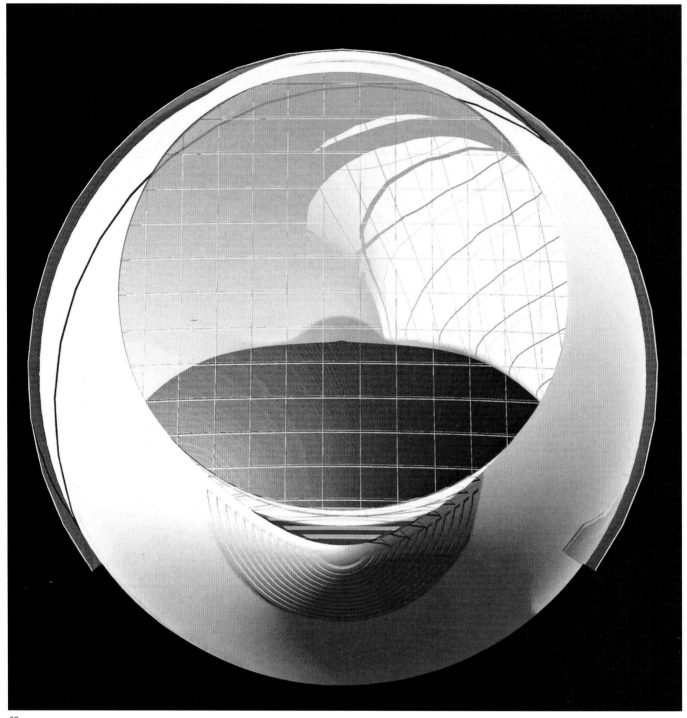

光眼

　　梦幻般的"光眼"组合在宫殿花园中形成了一种特殊的景象,如同悬挂的模型一样,受力是从边缘到中心的,然后沿着边缘传至支撑柱。"光眼"的边缘在结构中具有剪刀的功能,弧形玻璃边框镶嵌于混凝土边棱上。这个从地面突出的"光眼"构成了支撑体系的真正特征,并成为该原形组合体的结构胚胎。

32~35. "光眼"模型
32~35. Model- "Light eye"

"Light eye"

The fantastic "light eyes" creates a special scene in the palace garden. Like the suspension model, the force is distributed from the center to the edge, and then transmitted to the support column. The edge of the "light eye" functions as shearing in the structure, and the curved glass frame is embedded in the concrete edge. The protruding "light eye" constitutes the real characteristics of the supporting system, forming the structural embryo of the prototype.

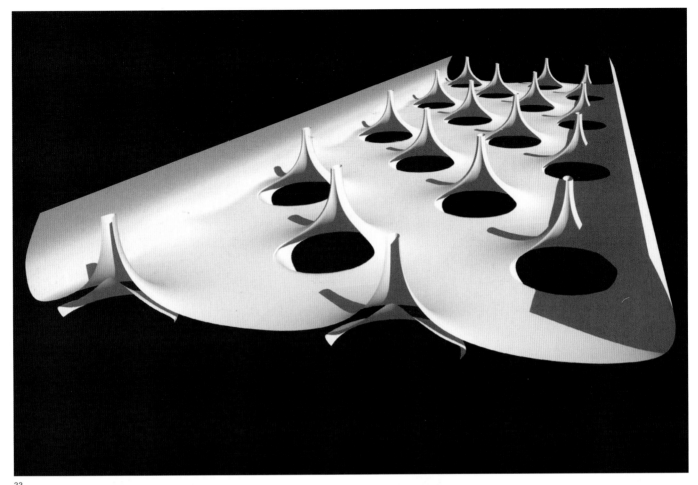
33

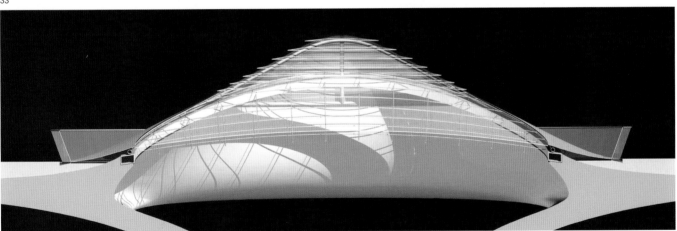
34

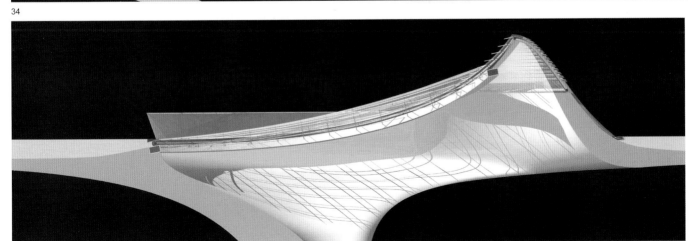
35

36

38

37

39

40

光线

光线在一个富有韵律的空间里扮演着重要的角色，斯图加特火车站的设计将成为新一代火车站的标志。当旅客在大厅里停留时，日光通过拱形玻璃壳均匀地进入大厅，即使在阴天也能得到舒适的光线。我们通过一个1：30的模型进行日光测试，结果显示：平均有5%的日光能直接到达内部，其中直接位于"光眼"下的部分能获得10%～15%的日光，在大约400m长的站台上每60m设有一个"光眼"，外部光线的变化使内部也能感觉到。由于两个相邻的站台每隔30m就有一个交叉的"光眼"，从而使光线均匀地进入地下站台。

36~37. 1：30 模型——自然光线分析
38. 1：30模型——夜晚灯光实验
39~40. 老车站室内改造
41. 地下车站效果图

36~37. 1：30 Model- Natural light analysis
38. 1：30Night lighting experiment
39~40. Interior renovation of old railway station
41. Station renderings

41

Light

The light played an important role in a rhythmic space. The design of the Stuttgart train station will be the emblem of a new generation of train stations. The daylight enters the hall evenly through the glass shell. The travelers are subject to comfortable natural lighting even on a cloudy day. A daylight test using the 1:30 model shows an average of 5% of the daylight reaches the internal directly, of which 10% - 15% is received directly under the "light eye". "Light eyes" are set up every 60m along the 400m platform, and the external light change is perceptible inside. Every 30 meters there is "light eye" alternatively from the two adjacent platforms, thus the natural light is distributed evenly on the platforms underground.

42

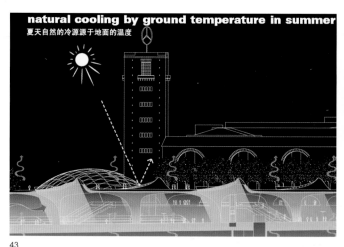

43

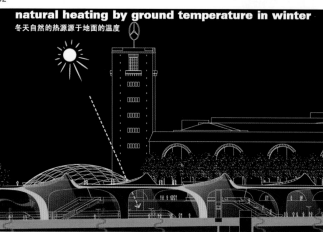

44

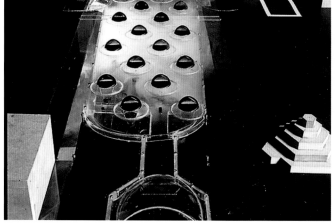

45

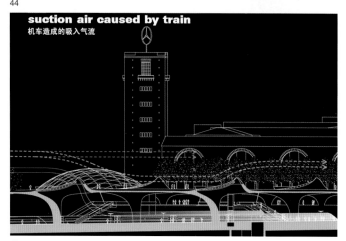

46

47

能源

斯图加特火车站是零能源建筑的代表，由于这种膜结构的形成，地道中的全年平均气温可控制在+10℃左右，在考虑气流的情况下计算出夏天的气温很少超过+20℃，冬天则很少低于0℃，由于高差的原因形成地下站台的壁炉效应。由天自然气候的原因，冬夏季将有不同温度的气流进入地下站台，这种冲击气流通过"光眼"在没有人工换气装置的情况下进入地下，其数据是每小时0.7倍的空气交换量（大约300 000m³/h）。在冬天可通过一个自动装置来阻挡低温气流的进入。

42. 能源分析——气流
43~44. 能源分析——温度
45. 能源分析——采光
46. 能源分析——气流
47. 1∶30模型实验
48. 模型照片

42. Energy Analysis - Air flow
43~44. Energy Analysis - temperature
45. Energy Analysis - lighting
46. Energy Analysis - Air flow
47. 1∶30 Model experiments
48. Model Photo

Energy

Stuttgart train station is a masterpiece of zero-energy building. Due to the formation of the membrane structure, the annual average temperature in the tunnel can be controlled at +10°C. Considering the airflow, the summer temperature rarely exceeds +20°C, while in winter it rarely drops below 0 °C as the height difference brings the fireplace effect on the underground platforms. The airflow entering underground platforms has different temperature depending on the seasonal conditions of the climate. Without artificial ventilation, the in-rush airflow through "light eye" can make up 0.7 times the amount of air exchange per hour (ca. 300000 m^3/h). An automatic device is used to block the low-temperature airflow in winter time.

49 鸟瞰图 AERIAL VIEW

50 室内效果图 INTERIOR RENDERING

参考文献：

英恩霍文、欧文迪克及其合伙人建筑师事务所著
《进化、生态和建筑》
英恩霍文、欧文迪克及其合伙人建筑师事务所著
《能源》
《建筑细部》（德文版）

References:

Ingenhoven Overdiek und Partner
Evolution Oekologie Architektur
Ingenhoven Overdiek und Partner
Energies
DETAIL Zeitschrift fuer Architektur

特别说明：

"重庆自然博物馆新馆"项目建筑师为五合国际建筑设计集团，主创设计师为卢卡斯（德国）；"合肥招商银行大厦"项目建筑师为五合国际建筑设计集团，主创建筑师为郭小平（德国）；"合肥MV广场"项目建筑师为五合国际建筑设计集团，主创建筑师为郭小平（德国）。

本书的理论和实践源于作者多年在英恩霍文建筑师事务所参加项目的总结和积累。
Book of theory and practice from the author for many years to participate in Ingenhoven Architects project summary and accumulation.

英恩霍文建筑师事务所网址/Ingenhoven Architects Website:
www.ingenhovenarchitects.com
德雅视界建筑师事务所网址/Ideal Architects Website:
www.ideals-design.com

项目解读建筑师简介：

克里斯多夫·英恩霍文
1978—1984年就读于德国亚琛工业大学
1980—1981年就读于杜塞尔多夫美术学院
克里斯多夫·英恩霍文自1985年起领导英恩霍文建筑师事务所（1985—2003年曾用名：英恩霍文、欧文迪克及其合伙人建筑师事务所），在主持事务所建筑设计任务的同时担任德国建筑评审委员、评审专家、国际招标项目评审官。

Introduction to Case Study Architects :

Christoph Ingenhoven
1978-1984 studied architecture at RWTH Aachen
1980-1981 studied architecture at Academy of Arts Düsseldorf under Prof. Hans Hollein
Since 1985 Christoph Ingenhoven is leading the Düsseldorf architectural office Ingenhoven Architects. (Until 2003 the company was named Ingenhoven Overdiek and Partner.) Besides his conceptual work Christoph Ingenhoven acts as jury member and expert in several competition processes and gives lectures all over the world.

作者简介：

郭小平
1964年出生于中国兰州，1999年毕业于德国杜塞尔多夫技术应用大学，获工学硕士，2011年为德国建筑师协会会员，德国注册建筑师。1994—2002年在德国英恩霍文建筑师事务所（Ingenhoven Architects）任建筑师，参加了汉莎航空公司总部办公楼、上海世茂国际广场、慕尼黑Upotown等节能建筑的设计和技术的发展。2002年回国曾就职于香港华艺建筑设计顾问有限公司、五合国际（5+Werkhart）、中国建筑技术集团有限公司建筑设计院，担任主创建筑师及设计总监职务，2010年创建德雅视界建筑师事务所并任设计总裁。

About the Author:

Xiaoping Guo
Was born in 1964 in Lanzhou, capital of northwest China's Gansu Province. He graduated from the Technical University of Applied Sciences Dusseldorf, Germany, and obtained a Master of Engineering. He is a member of the German Association of Architects and a registered architect in Germany since 2001. From 1994 to 2002, Xiaoping worked as architect for Ingenhoven & Partner in Germany, participated in the design of various projects including the Lufthansa headquarter office building, Shanghai Shimao International Plaza, Munich Uptown and other energy efficient building design and technology development.Back to China since 2002, he has been chief architect for Hong Kong Hua-Yi Building Design Consultants Ltd., 5 + Werkhart design director of Architectural Design Institute at China Construction Technology Group Co., Ltd.; and from 2010 he is the president (chief executive officer) of Ideal Architects.

英文译文：

王蓓敏

1975年出生于中国上海，初中进入上海外国语大学附属外国语学校学习德语。1996年赴德留学，于杜塞尔多夫市海涅大学专攻日耳曼语言文学及英国文学，并以优异成绩获得双语硕士学位。曾与本书作者共事于上海世茂国际广场的项目，担任德方设计团队的专职翻译。2003年移居美国，现居上海和美国奥斯汀市。

英文校对：

威廉·罗本

先后毕业于美国加州大学伯克利分校（景观建筑设计学）和美国哈佛大学设计学院研究院，获得城市规划设计硕士学位。美国加州注册景观建筑师。历年的参与作品包括美国加利福尼亚州的丰田美国总部、好莱坞环球影城、美国伊利诺斯州的麦当劳全球总部、派拉蒙电影公司中国制片厂、韩国仁川度假村、日本大阪亚太贸易中心等。现任美国加州的罗本/赫尔曼设计公司的合伙人兼首席设计总监。

中文校对：

刘芳

封面设计：

季娜

关澍

版面设计：

葛峰

English Translator:

Beimin Wang

Born in 1975 in Shanghai, China, Beimin Wang entered Shanghai Foreign Language School to study German in her young age. She went Heinrich Heine University in Düsseldorf, Germany, in 1996, major in Germanic Philology and English Literature, and obtained a bilingual master's degree with honor. She worked with the author on the project of Shanghai Shimao International Plaza as the translator on duty for the German design team. Moved to the United States in 2003, she now lives in Austin, U.S. and in Shanghai, China.

English Proofreader:

William Rabben

Graduated from University of California at Berkeley with a BA (Honors) in Landscape Architecture, William Rabben also holds a Master of Urban Planning from the Harvard University Graduate School of Design. He is a registered landscape architect in the state of California. Over the years, he worked on various important projects including: Toyota headquarters in California, Hollywood Universal Studios, McDonald global headquarters in Illinois, Paramount Movie Studios in China, Incheon Urban Resort in Korea, and Asia Pacific Trade Center in Osaka, Japan. Currently, he is the Chief Design Officer of Rabben Herman Design Office, Newport Beach, CA.

Chinese Proofreader:

Fang Liu

Cover Design:

Na Ji

Shu Guan

Layout Design:

Feng Ge

图书在版编目(CIP)数据

走向未来建筑 / 郭小平编著 . – 武汉：华中科技大学出版社, 2014.9
ISBN 978-7-5680-0330-8

Ⅰ.①走… Ⅱ.①郭… Ⅲ.①生态建筑 – 建筑设计 Ⅳ.①TU201.5

中国版本图书馆CIP数据核字(2014)第183256号

走向未来建筑

郭小平 编著

出版发行：华中科技大学出版社（中国·武汉）
地　　址：武汉市武昌珞喻路1037号（邮编:430074）
出 版 人：阮海洪

责任编辑：刘锐桢　　　　　　　　　　　　　　　　　　　　责任监印：秦　英
责任校对：杨　睿　　　　　　　　　　　　　　　　　　　　装帧设计：张　靖

印　　刷：北京利丰雅高长城印刷有限公司
开　　本：965 mm×1270 mm　1/16
印　　张：15.25
字　　数：121千字
版　　次：2014年9月第1版第1次印刷
定　　价：248.00元

投稿热线：(010)64155588-8815
本书若有印装质量问题，请向出版社营销中心调换
全国免费服务热线：400-6679-118 竭诚为您服务
版权所有　侵权必究